STRUCTURE ELUCIDATION
BY NMR
IN ORGANIC CHEMISTRY
A Practical Guide

Third revised edition

EBERHARD BREITMAIER

University of Bonn, Germany

JOHN WILEY & SONS, LTD

Other Wiley Editorial Offices

John Wiley & Sons Inc., 111 River Street, Hoboken, NJ 07030, USA

Jossey-Bass, 989 Market Street, San Francisco, CA 94103-1741, USA

Wiley-VCH Verlag GmbH, Boschstr. 12, D-69469 Weinheim, Germany

John Wiley & Sons Australia Ltd, 33 Park Road, Milton, Queensland 4064, Australia

John Wiley & Sons (Asia) Pte Ltd, 2 Clementi Loop #02-01, Jin Xing Distripark, Singapore 129809

John Wiley & Sons Canada Ltd, 22 Worcester Road, Etobicoke, Ontario, Canada M9W 1L1

British Library Cataloguing in Publication Data

A catalogue record for this book is available from the British Library

ISBN 0 470 85006 X (Cloth)
ISBN 0 470 85007 8 (Paper)

Produced from camera ready copy supplied by the author.
Printed and bound in Great Britain by Antony Rowe Ltd, Chippenham, Wiltshire.
This book is printed on acid-free paper responsibly manufactured from sustainable forestry in which at least
two trees are planted for each one used for paper production.

The cover shows the ^{13}C NMR spectrum of α- and β-D-xylopyranose at mutarotational equilibrium (35% α,
65% β, in deuterium oxide, 100 MHz, ^{1}H broadband decoupling) with the CC INADEQUATE contour plot.
An interpretation of the plot according to principles described in Section 2.2.7 gives the CC bonds of the two
isomers and confirms the assignment of the signals in Table 2.12.

Books are to be returned on or before the last date below.

LIBREX —

Contents

PREFACE

Virtually, all students of chemistry, biochemistry, pharmacy and related subjects learn how to deduce molecular structures from nuclear magnetic resonance (NMR) spectra. Undergraduate examinations routinely set problems using NMR spectra, and masters' and doctoral theses describing novel synthetic or natural products provide many examples of how powerful NMR has become in structure elucidation. Existing texts on NMR spectroscopy generally deal with the physical background of the newer and older techniques as well as the relationships between NMR parameters and chemical structures. Very few, however, convey the know-how of structure determination using NMR, namely the strategy and methodology by which molecular structures are deduced from NMR spectra.

This book, based on many lectures and seminars, attempts to provide advanced undergraduates and graduate students with a systematic, readable and inexpensive introduction to the methods of structure determination by NMR. Chapter 1 starts with a deliberately concise survey of the basic terms, parameters and techniques dealt with in detail in other books, which cover the basic principles of NMR, pulse sequences as well as theoretical aspects of chemical shifts and spin-spin coupling, and which this workbook is not intended to replace. An introduction to basic strategies and tactics of structure determination using one- and two-dimensional NMR methods then follows in Chapter 2. Here, the emphasis is always on how spectra and associated parameters can be used to identify structural fragments. This chapter presents those topics that are essential for the identification of compounds or for solving structures, including atom connectivities, relative configuration and conformation, intra- and intermolecular interactions and, in some cases, molecular dynamics. Following the principle of 'learning by doing', Chapter 3 presents a series of case studies, providing spectroscopic details of 55 compounds that illustrate typical applications of NMR techniques in the structural characterisation of both synthetic and natural products. The level of difficulty, the sophistication of the methodology required increases from question to question, so that all readers will be able to find material suited to their knowledge and ability. One can work independently, solve the problem from the spectra and check the result in the formula index, or follow the detailed solutions given in Chapter 4. The spectroscopic details are presented in a way that makes the maximum possible information available at a glance, requiring minimal page turning. Chemical shifts and coupling constants do not have to be read off from scales but are presented numerically, allowing the reader to concentrate directly on problem solving without the need for tedious routine work.

Actual methods of two dimensional NMR such as some inverse techniques of heteronuclear shift correlation experiments (HMQC, HSQC, HMBC), proton shift correlations (TOCSY) and two-dimensional detection of nuclear Overhauser effects (ROESY) are illustrated in Chapter 2 of this edition. New problems are added in Chapter 3 and 4 not only to replace some of the former ones but also in order to improve the quality and to demonstrate some applications of the actual methods shown in Chapter 2. All formulae have been redrawn using new software; all spectra have been scanned into the data file and the layout has been optimized. My thanks must go to Dr. Rudolf Hartmann for recording some of the two-dimensional NMR experiments, to Klaus Rotscheidt for scanning and his assistance in electronic editing, and especially to Julia Wade for having translated the original German text for the first English edition of this book.

Eberhard Breitmaier

SYMBOLS AND ABBREVIATIONS

APT: Attached proton test, a modification of the J-modulated spin-echo experiment to determine CH multiplicities, a less sensitive alternative to DEPT

CH COLOC: Correlation *via* long-range CH coupling, detects CH connectivities through two or three (more in a few cases) bonds in the CH COSY format, permits localisation of carbon nuclei two or three bonds apart from an individual proton

COSY: Correlated spectroscopy, two-dimensional shift correlations *via* spin-spin coupling, homonuclear (e.g. HH) or heteronuclear (e.g. CH)

CH COSY: Correlation *via* one-bond CH coupling, also referred to as HETCOR (heteronuclear shift correlation), provides carbon-13- and proton shifts of nuclei in CH bonds as cross signals in a δ_C *versus* δ_H diagram, assigns all CH bonds of the sample

HH COSY: Correlation *via* HH coupling which has square symmetry because of equal shift scales in both dimensions (δ_H *versus* δ_H) provides all detectable HH connectivities of the sample

CW: Continuous wave or frequency sweep, the older, less sensitive, more time consuming basic technique of NMR detection

DEPT: Distortionless enhancement by polarisation transfer, differentiation between CH, CH_2 and CH_3 by positive (CH, CH_3) or negative (CH_2) signal amplitudes, using improved sensitivity of polarisation transfer

FID: Free induction decay, decay of the induction (transverse magnetisation) back to equilibrium (transverse magnetisation zero) due to spin-spin relaxation, following excitation of a nuclear spin by a radio frequency pulse, in a way which is free from the influence of the radiofrequency field; this signal (time-domain) is Fourier-transformed to the FT NMR spectrum (frequency domain)

FT NMR: Fourier transform NMR, the newer and more sensitive, less time consuming basic technique of NMR detection, almost exclusively used

INADEQUATE: Incredible natural abundance double quantum transfer experiment, segregates AB or AX systems due to homonuclear one-bond couplings of less abundant nuclei, e.g. $^{13}C-^{13}C$; CC INADEQUATE detects CC bonds (carbon skeleton) present in the sample

HMBC: Heteronuclear multiple bond correlation, inverse CH correlation *via* long-range CH coupling, same format and information as described for (^{13}C detected) CH COLOC but much more sensitive (therefore less time-consuming) because of ^1H detection

HMQC: Heteronuclear multiple quantum coherence, e.g. inverse CH correlation *via* one-bond carbon proton-coupling, same format and information as described for (^{13}C detected) CH COSY but much more sensitive (therefore less time-consuming) because of ^1H detection

HSQC: Heteronuclear single quantum coherence, e.g. inverse C*H* correlation *via* one-bond coupling providing the same result as HMQC but using an alternative pulse sequence

NOE: Nuclear Overhauser effect, change of signal intensities (integrals) during decoupling experiments decreasing with spatial distance of nuclei

NOESY: Nuclear Overhauser effect spectroscopy, detection of NOE in the *HH* COSY square format, traces out closely spaced protons in larger molecules

ROESY: Rotating frame NOESY, detection of NOE in the *HH* COSY format with suppressed spin-diffusion, detects closely spaced protons also in smaller molecules

TOCSY: Total correlation spectroscopy, in the homonuclear COSY format, e.g. *HH* TOCSY traces out all proton-proton connectivities of a partial structure in addition to the connectivities (^2J, ^3J, ^4J, ^5J) as detected by *HH* COSY

J or ^1J: nuclear spin-spin coupling constant (in Hz) through *one* bond (one-bond coupling)

^2J, ^3J, ^4J, ^5J: nuclear spin-spin coupling constant (in Hz) through *two*, *three*, *four* and *five* bonds (*geminal*, *vicinal*, *longer-range* couplings)

Multiplet abbreviations:

S, s	:	singlet
D, d	:	doublet
T, t	:	triplet
Q, q	:	quartet
Qui, qui	:	quintet
Sxt, sxt	:	sextet
Sep, sep	:	septet
o	:	overlapping
b	:	broad

| Capital letters: | multiplets which are the result of coupling through one bond |
| Lower case letters: | multiplets which are the result of coupling through more bonds than one |

δ_H, δ_C, δ_N :   *Proton, carbon-13 and nitrogen-15 chemical shifts*

Contrary to IUPAC conventions, chemical shifts δ in this book are scaled in ppm in the spectra, thus enabling the reader to differentiate at all times between shift values (ppm) and coupling constants (Hz); ppm (parts per million) is in this case the ratio of two frequencies of different orders of magnitude, Hz / MHz = 1 : 10^6 without physical dimension

Italicised data and multiplet abbreviations refer to ^1H in this book

1 SHORT INTRODUCTION TO BASIC PRINCIPLES AND METHODS

1.1 Chemical shift

Chemical shift relates the Larmor frequency of a nuclear spin to its chemical environment [1-3]. The Larmor frequency is the precession frequency v_0 of a nuclear spin in a static magnetic field (Fig. 1.1). This frequency is proportional to the flux density B_0 of the magnetic field (v_0/B_0 = const.) [1-3]. It is convenient to reference the chemical shift to a standard such as tetramethylsilane [TMS, $(CH_3)_4Si$] rather than to the proton H^+. Thus, a frequency difference (Hz) is measured for a proton or a carbon-13 nucleus of a sample from the ^1H or ^{13}C resonance of TMS. This value is divided by the absolute value of the Larmor frequency of the standard (e.g. 400 MHz for the protons and 100 MHz for the carbon-13 nuclei of TMS when using a 400 MHz spectrometer), which itself is proportional to the strength B_0 of the magnetic field. The chemical shift is therefore given in parts per million (*ppm*, δ scale, δ_H for protons, δ_C for carbon-13 nuclei), because a frequency difference in Hz is divided by a frequency in MHz, these values being in a proportion of $1:10^6$.

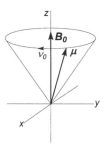

Figure 1.1. Nuclear precession: nuclear charge and nuclear spin give rise to a magnetic moment of nuclei such as protons and carbon-13. The vector μ of the magnetic moment precesses in a static magnetic field with the Larmor frequency v_0 about the direction of the magnetic flux density vector B_0

Chemical shift is principally caused by the electrons in the molecule having a *shielding effect* on the nuclear spin. More precisely, the electrons cause a *shielding field* which opposes the external magnetic field: the precession frequency of the nuclear spin (and in turn the size of its chemical shift) is therefore reduced. An atomic nucleus (e.g. a proton) whose shift is reduced is said to be *shielded* (high shielding field); an atom whose shift is increased is said to be *deshielded* (low shielding field). This is illustrated in Fig. 1.2 which also shows that NMR spectra are presented with chemical shift and frequency decreasing from left to right.

1.2 Spin-spin coupling and coupling constants

Indirect or *scalar coupling* [1-3] of nuclear spins through covalent bonds causes the splitting of NMR signals into multiplets in high-resolution NMR spectroscopy in the solution state. The direct or *dipolar coupling* between nuclear spins through space is only observed for solid or liquid crystalline samples. In a normal solution such coupling is cancelled out by molecular motion.

The *coupling constant* for first-order spectra (see Section 1.4) is the frequency difference J in Hz between two multiplet lines. Unlike chemical shift, the frequency value of a coupling constant does not depend on the strength of the magnetic field. In high-resolution NMR a distinction is made between coupling through one bond (^1J or simply J, *one-bond* couplings) and coupling through several bonds, e.g. two bonds (^2J, *geminal* couplings), three bonds (^3J, *vicinal* couplings) or four or five bonds (^4J and ^5J, long-range couplings). For example, the CH_2 and CH_3 protons of the ethyl group in ethyldichloroacetate (Fig. 1.2) are separated by three bonds; their (*vicinal*) coupling constant is $^3J = 7\ Hz$.

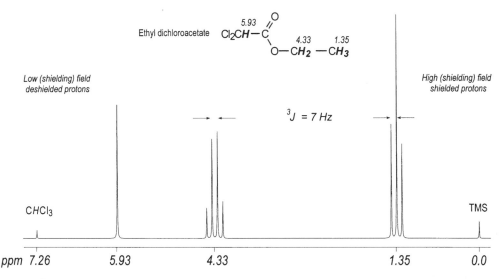

Figure 1.2. ^1H NMR spectrum of ethyl dichloroacetate ($CDCl_3$, 25 °C, 80 MHz). The proton of the $CHCl_2$ group is less shielded (more strongly deshielded) in comparison with the protons of the CH_2 and CH_3 residues

1.3   Signal multiplicity (multiplets)

The *signal multiplicity* in first-order spectra (see Section 1.4) is the extent to which an NMR signal is split as a result of spin-spin coupling [1-3]. Signals which show no splitting are denoted as *singlets* (*s*). Those with two, three, four, five, six or seven lines are known as *doublets* (*d*), *triplets* (*t*), *quartets* (*q*, Figs 1.2 and 1.3), *quintets* (*qui*), *sextets* (*sxt*) and *septets* (*sep*), respectively, but only where the lines of the multiplet signal are of equal distance apart, and the one coupling constant is therefore shared by them all. Where two or three different coupling constants produce a multiplet, this is referred to as a two- or three-fold multiplet, respectively, e.g. a *doublet of doublets* (*dd*, Fig. 1.3), or a *doublet of doublets of doublets* (*ddd*, Fig. 1.3). If both coupling constants of a doublet of doublets are sufficiently similar ($J_1 \sim J_2$), the middle signals overlap, thus generating a *'pseudotriplet'* (*'t'*, Fig. 1.3).

The ^1H NMR spectrum of ethyl dichloroacetate (Fig. 1.2), as an example, displays a triplet for the CH_3 group (*two vicinal H*), a quartet for the OCH_2 group (three *vicinal H*) and a singlet for the $CHCl_2$ fragment (no *vicinal H* for coupling).

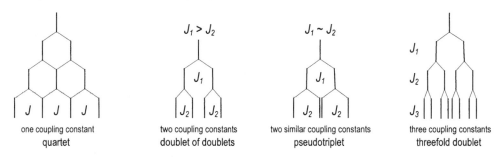

one coupling constant
quartet

two coupling constants
doublet of doublets

two similar coupling constants
pseudotriplet

three coupling constants
threefold doublet

Figure 1.3. Quartet, doublet of doublets, pseudotriplet and threefold doublet (doublet of doublets of doublets)

1.4   Spectra of first and higher order

First-order spectra (*multiplets*) are observed when the coupling constant is small compared with the frequency difference of chemical shifts between the coupling nuclei [2,3]. This is referred to as an A_mX_n spin system, where nucleus A has the smaller and nucleus X has the considerably larger chemical shift. An AX system (Fig. 1.4) consists of an A doublet and an X doublet with the common coupling constant J_{AX}. The chemical shifts are measured from the centres of each doublet to the reference resonance.

Figure 1.4. Two-spin system of type AX with a chemical shift difference which is large compared with the coupling constants (schematic)

n = 0	Singlet											1							
1	Doublet									1	:	1							
2	Triplet							1	:	2	:	1							
3	Quartet					1	:	3	:	3	:	1							
4	Quintet			1	:	4	:	6	:	4	:	1							
5	Sextet		1	:	5	:	10	:	10	:	5	:	1						
6	Septet	1	:	6	:	15	:	20	:	15	:	6	:	1					

Figure 1.5. Relative intensities of first-order multiplets (Pascal triangle)

Multiplicity rules apply for first-order spectra (A_mX_n systems): When coupled, n_X nuclei of an element X with nuclear spin quantum number $I_X = \frac{1}{2}$ produce a splitting of the A signal into $n_X + 1$ lines; the relative intensities of the individual lines of a first-order multiplet are given by the coefficients of the Pascal triangle (Fig. 1.5). The protons of the ethyl group of ethyl dichloroacetate (Fig. 1.2) as examples give rise to an A_3X_2 system with the coupling constant $^3J_{AX} = 7\ Hz$; the A protons (with smaller shift) are split into a triplet (two *vicinal* protons X, $n_X + 1 = 3$); the X protons appear as a quartet because of three *vicinal* A protons ($n_A + 1 = 4$). In general, for a given number, n_X, of coupled nuclear spins of spin quantum number I_X, the A signal will be split into ($2\,n_X I_X + 1$) multiplet lines (e.g. Fig. 1.9).

Spectra of greater complexity may occur for systems where the coupling constant is of similar magnitude to the chemical shift difference between the coupled nuclei. Such a case is referred to as an A_mB_n system, where nucleus A has the smaller and nucleus B the larger chemical shift.

An AB system (Fig. 1.6) consists, for example, of an A doublet and a B doublet with the common coupling constant J_{AB}, where the external signal of both doublets is attenuated and the internal signal is enhanced. This is referred to as an *AB effect,* a 'roofing' symmetric to the centre of the AB system [2]; 'roofing' is frequently observed in proton NMR spectra, even in practically first order spectra (Fig. 1.2, ethyl quartet and triplet). The chemical shifts v_A and v_B are displaced from the centres of the two doublets, approaching the frequencies of the more intense inner signals.

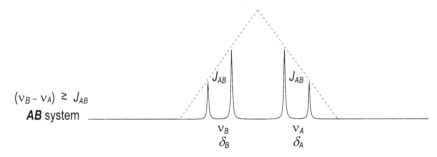

Figure 1.6. Two-spin system of type *AB* with a small chemical shift difference compared to the coupling constant (schematic)

1.5   Chemical and magnetic equivalence

Chemical equivalence: atomic nuclei in the same chemical environment are chemically equivalent and thus show the same chemical shift. The 2,2'- and 3,3'-protons of a 1,4-disubstituted benzene ring, for example, are chemically equivalent because of molecular symmetry.

Magnetic equivalence: chemically equivalent nuclei are magnetically equivalent if they display the same coupling constants with all other nuclear spins of the molecule [2,3]. For example, the 2,2'-

(AA′) and 3,3'-*(X,X′)* protons of a 1,4-disubstituted benzene ring such as 4-nitroanisole are not magnetically equivalent, because the 2-proton *A* shows an *ortho* coupling with the 3-proton *X* ($^3J =$ *7-8 Hz*), but displays a different *para* coupling with the 3'-proton *X′* ($^5J = 0.5$ - *1 Hz*). This is therefore referred to as an *AA′XX′* system (e.g. Fig. 2.6 **c**) but not as an A_2X_2 or an *(AX)$_2$* system. The ^1H NMR spectrum in such a case can never be first-order, and the multiplicity rules do not apply. The methyl protons in ethyl dichloroacetate (Fig. 1.2), however, are chemically and magnetically equivalent because the $^3J_{HH}$ coupling constants depend on the geometric relations with the CH_2 protons and these average to the same for all CH_3 protons due to rotation about the CC single bond; they are the A_3 part of an A_3X_2 system characterising an ethoxy group ($CH^A_3 - CH^X_2 - O-$) in ^1H NMR.

1.6   Fourier transform (FT) NMR spectra

There are two basic techniques for recording high-resolution NMR spectra [2-6]. In the older *CW technique,* the frequency or field appropriate for the chemical shift range of the nucleus (usually ^1H) is swept by a continuously increasing (or decreasing) radio-frequency. The duration of the sweep is long, typically *2 Hz/s,* or *500 s* for a sweep of *1000 Hz,* corresponding to *10 ppm* in 100 MHz *proton* NMR spectra. This monochromatic excitation therefore takes a long time to record.

In the *FT technique,* the whole of the Larmor frequency range of the observed nucleus is excited by a radiofrequency pulse. This causes transverse magnetisation to build up in the sample. Once excitation stops, the transverse magnetisation decays exponentially with the time constant T_2 of spin-spin relaxation provided the field is perfectly homogeneous. In the case of a one-spin system, the corresponding NMR signal is observed as an exponentially decaying alternating voltage (*free induction decay,* FID); multi-spin systems produce an interference of several exponentially decaying alternating voltages, the *pulse interferogram* (Fig. 1.7).The frequency of each alternating voltage is the difference between the individual Larmor frequency of one specific kind of nucleus and the frequency of the exciting pulse. The Fourier transformation (FT) of the pulse interferogram produces the Larmor frequency spectrum; this is the *FT NMR spectrum* of the type of nucleus being observed. Fourier transformation involving the calculation of all Larmor frequencies contributing to the interferogram is performed with the help of a computer within a time of less than 1 s.

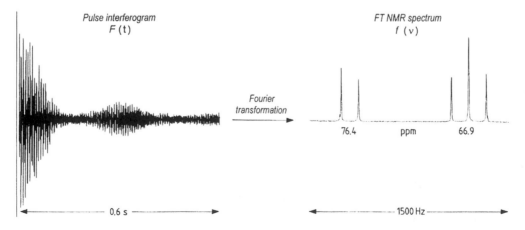

Figure 1.7. Pulse interferogram and FT ^{13}C NMR spectrum of glycerol, (DOCH_2)$_2$CHOD, [D$_2$O, 25 °C, 100 MHz]

The main advantage of the *FT* technique is the short time required for the procedure (about 1 s per interferogram). Within a short time a large number of individual interferograms can be accumulated, thus averaging out electronic noise (*FID accumulation*), and making the *FT* method the preferred approach for less sensitive NMR probes involving isotopes of low natural abundance (^{13}C, ^{15}N). All of the spectra in this book with the exception of those in Figs. 1.8, 2.19 and 2.25 are *FT* NMR spectra.

1.7   Spin decoupling

Spin decoupling (*double resonance*) [2,3,5,6] is an NMR technique in which, to take the simplest example, an *AX* system, the splitting of the *A* signal due to J_{AX} coupling is removed if the sample is irradiated strong enough by a second radiofrequency which resonates with the Larmor frequency of the *X* nucleus. The *A* signal then appears as a singlet; at the position of the *X* signal interference is observed between the *X* Larmor frequency and the decoupling frequency. If the *A* and *X* nuclei are the same isotope (e.g. protons), this is referred to as *selective homonuclear decoupling*. If *A* and *X* are different, e.g. carbon-13 and protons, then it is referred to as *heteronuclear decoupling*.

Figure 1.8 illustrates homonuclear decoupling experiments with the *CH* protons of 3-aminoacrolein. These give rise to an *AMX* system (Fig. 1.8**a**). Decoupling of the aldehyde proton *X* (Fig. 1.8**b**) simplifies the NMR spectrum to an *AM* system ($^{3}J_{AM} = 12.5$ *Hz*); decoupling of the *M* proton (Fig. 1.8**c**) simplifies to an *AX* system ($^{3}J_{AX} = 9$ *Hz*). These experiments reveal the connectivities of the protons within the molecule.

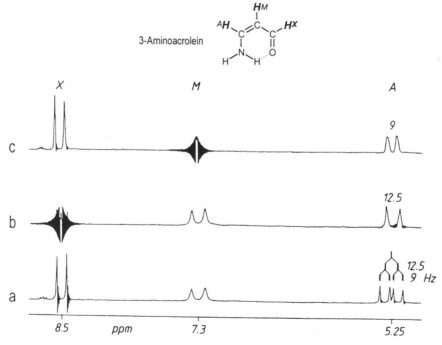

Figure 1.8. Homonuclear decoupling of the *CH* protons of 3-aminoacrolein (CD₃OD, 25 °C, 90 MHz). (**a**) ^{1}H NMR spectrum; (**b**) decoupling at $\delta_H = 8.5$; (**c**) decoupling at $\delta_H = 7.3$. At the position of the decoupled signal in (**b**) and (**c**) interference beats are observed because of the superposition of the two very similar frequencies

In ^{13}C NMR spectroscopy, three kinds of heteronuclear spin decoupling are used [5,6]. In *proton broadband decoupling* of ^{13}C NMR spectra, decoupling is carried out unselectively across a frequency range which encompasses the whole range of the proton shifts. The spectrum then displays up to *n* singlet signals for the *n* non-equivalent C atoms of the molecule.

Figures 1.9**a** and **b** demonstrate the effect of proton broadband decoupling in the ^{13}C NMR spectrum of a mixture of ethanol and hexadeuterioethanol. The *CH₃* and *CH₂* signals of ethanol appear as intense singlets upon proton broadband decoupling while the CD₃ and CD₂ resonances of the deuteriated compound still display their septet and quintet fine structure; deuterium nuclei are not affected by ^{1}H decoupling because their Larmor frequencies are far removed from those of protons; further, the nuclear spin quantum number of deuterium is $I_D = 1$; in keeping with the general multiplicity rule ($2\,n_X\,I_X + 1$, Section 1.4), triplets, quintets and septets are observed for CD, CD₂ and CD₃ groups, respectively. The relative intensities in these multiplets do not follow Pascal's triangle (1:1:1 triplet for CD; 1:3:4:3:1 quintet for CD₂; 1:3:6:7:6:3:1 septet for CD₃).

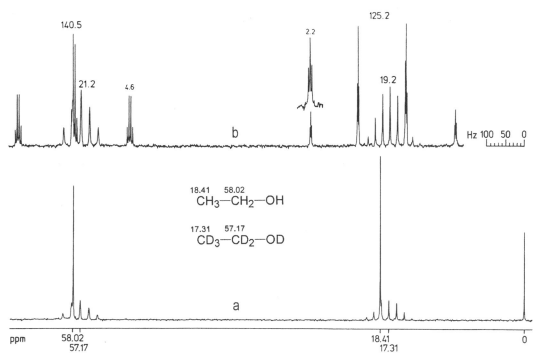

Figure 1.9. ^{13}C NMR spectra of a mixture of ethanol and hexadeuterioethanol [27:75 v/v, 25 °C, 20 MHz]. (**a**) ^{1}H broadband decoupled; (**b**) without decoupling. The deuterium isotope effect $\delta_{CH} - \delta_{CD}$ on ^{13}C chemical shifts is 1.1 and 0.85 ppm for methyl and methylene carbon nuclei, respectively

In *selective proton decoupling* of ^{13}C NMR spectra, decoupling is performed at the precession frequency of a specific proton. As a result, a singlet only is observed for the attached C atom. *Off-resonance* conditions apply to the other C atoms. For these the individual lines of the C*H* multiplets move closer together, and the relative intensities of the multiplet lines change from those given by the Pascal triangle; external signals are attenuated whereas internal signals are enhanced.

Selective ^1H *decoupling* of ^{13}C NMR spectra was used for assignment of the CH connectivities (CH bonds) before the much more efficient two-dimensional CH shift correlation experiments (see Section 2.2.8) became routine. *Off-resonance decoupling* of the protons was helpful in determining CH multiplicities before better methods became available (see Section 2.2.2). In *pulsed* or *gated decoupling* of protons (broadband decoupling only *between* FIDs), coupled ^{13}C NMR spectra are obtained in which the CH multiplets are enhanced by the *nuclear Overhauser effect* (NOE, see Section 1.8). This method is used when CH coupling constants are required for structure analysis because it enhances the multiplets of carbon nuclei attached to protons; the signals of quaternary carbons two bonds apart from a proton are also significantly enhanced. Figure 1.10 demonstrates this for the carbon nuclei in the 4,6-positions of 2,4,6-trichloropyrimidine.

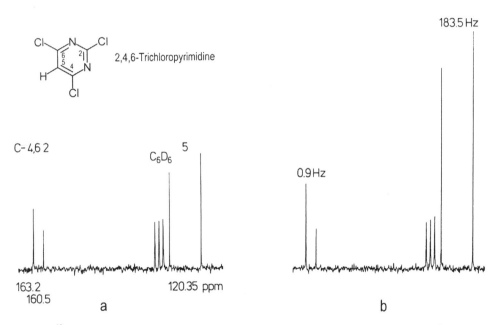

Figure 1.10. ^{13}C NMR spectra of 2,4,6-trichloropyrimidine [C_6D_6, 75% v/v 25 °C, 20 MHz]. (**a**) ^{13}C NMR spectrum without proton decoupling; (**b**) NOE enhanced coupled ^{13}C NMR spectrum (gated decoupling)

1.8   Nuclear Overhauser effect

The *nuclear Overhauser effect* [2,3] (NOE, also an abbreviation for nuclear Overhauser enhancement) causes the change in intensity (increase or decrease) during decoupling experiments. The maximum possible NOE in high-resolution NMR of solutions depends on the gyromagnetic ratio of the coupled nuclei. Thus, in the homonuclear case such as proton-proton coupling, the NOE is much less than 0.5, whereas in the most frequently used heteronuclear example, proton decoupling of ^{13}C NMR spectra, it may reach 1.988. Instead of the expected signal intensity of 1, the net result is to increase the signal intensity threefold (1 + 1.988). In proton broadband and gated decoupling of ^{13}C NMR spectra, NOE enhancement of signals by a factor of as much as 2 is routine, as was shown in Figs 1.9 and 1.10.

Quantitative analysis of mixtures is achieved by evaluating the integral steps of ^1H NMR spectra. This is demonstrated in Fig. 1.11a for 2,4-pentanedione (acetylacetone) which occurs as an equilibrium mixture of 87 % enol and 13 % diketone.

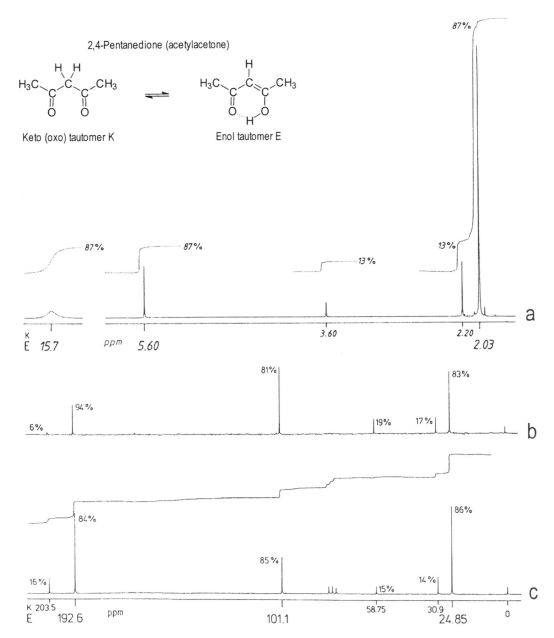

Figure 1.11. NMR analysis of the keto-enol tautomerism of 2,4-pentanedione [CDCl$_3$, 50% v/v, 25 °C, 60 MHz for ^1H, 20 MHz for ^{13}C]. (**a**) ^1H *NMR* spectrum with integrals [result: keto : enol = 13 : 87]; (**b**) ^1H broadband decoupled ^{13}C NMR spectrum; (**c**) ^{13}C NMR spectrum obtained by inverse gated ^1H decoupling with integrals [result: keto : enol = 15 : 85 (± 1)]

A similar evaluation of the ^{13}C integrals in ^{1}H broadband decoupled ^{13}C NMR spectra fails in most cases because signal intensities are influenced by nuclear Overhauser enhancements and relaxation times and these are usually specific for each individual carbon nucleus within a molecule. As a result, deviations are large (81 - 93 % enol) if the keto-enol equilibrium of 2,4-pentanedione is analysed by means of the integrals in the ^{1}H broadband decoupled ^{13}C NMR spectrum (Fig. 1.11**b**). *Inverse gated decoupling* [2,3], involving proton broadband decoupling only during the FIDs, helps to solve the problem. This technique provides ^{1}H broadband decoupled ^{13}C NMR spectra with suppressed nuclear Overhauser effect so that signal intensities can be compared and keto-enol tautomerism of 2,4-pentanedione, for example, is analysed more precisely as shown in Fig. 1.11**c**.

1.9   Relaxation, relaxation times

Relaxation [2,3,6] refers to all processes which regenerate the Boltzmann distribution of nuclear spins on their precession states and the resulting equilibrium magnetisation along the static magnetic field. Relaxation also destroys the transverse magnetisation arising from phase coherence of nuclear spins built up upon NMR excitation.

Spin-lattice relaxation is the steady (exponential) build-up or regeneration of the Boltzmann distribution (equilibrium magnetisation) of nuclear spins in the static magnetic field. The lattice is the molecular environment of the nuclear spin with which energy is exchanged.

The *spin-lattice relaxation time*, T_1, is the time constant for spin-lattice relaxation which is specific for every nuclear spin. In FT NMR spectroscopy the spin-lattice relaxation must 'keep pace' with the exciting pulses. If the sequence of pulses is too rapid, e.g. faster than $3T_{1max}$ of the 'slowest' C atom of a molecule in carbon-13 resonance, a decrease in signal intensity is observed for the 'slow' C atom due to the spin-lattice relaxation getting 'out of step'. For this reason, quaternary C atoms can be recognised in carbon-13 NMR spectra by their weak signals.

Spin-spin relaxation is the steady decay of transverse magnetisation (phase coherence of nuclear spins) produced by the NMR excitation where there is perfect homogeneity of the magnetic field. It is evident in the shape of the FID (*free induction decay*), as the exponential decay to zero of the transverse magnetisation produced in the pulsed NMR experiment. The Fourier transformation of the FID signal (time domain) gives the FT NMR spectrum (frequency domain, Fig. 1.7).

The *spin-spin relaxation time*, T_2, is the time constant for spin-spin relaxation which is also specific for every nuclear spin (approximately the time constant of FID). For small- to medium-sized molecules in solution $T_2 \sim T_1$. The value of T_2 of a nucleus determines the width of the appropriate NMR signal at half-height ('half-width') according to the uncertainty relationship. The smaller is T_2, the broader is the signal. The more rapid is the molecular motion, the larger are the values of T_1 and T_2 and the sharper are the signals (*'motional narrowing'*). This rule applies to small- and medium-sized molecules of the type most common in organic chemistry.

Chemical shifts and coupling constants reveal the static structure of a molecule; relaxation times reflect molecular dynamics.

2 RECOGNITION OF STRUCTURAL FRAGMENTS BY NMR

2.1 Functional groups

2.1.2 ^1H Chemical Shifts

Many functional groups can be identified conclusively by their ^1H chemical shifts [1-3]. Important examples are listed in Table 2.1, where the ranges for the proton shifts are shown in decreasing sequence: aldehydes ($\delta_H = 9.5 - 10.5$), acetals ($\delta_H = 4.5 - 6$), alkoxy ($\delta_H = 4 - 5.5$) and methoxy functions ($\delta_H = 3.5 - 4$), N-methyl groups ($\delta_H = 3 - 3.5$) and methyl residues attached to double bonds such as C=C or C=X (X = N, O, S) or to aromatic and heteroaromatic skeletons ($\delta_H = 1.8 - 2.5$).

Table 2.1. ^1H chemical shift ranges for organic compounds

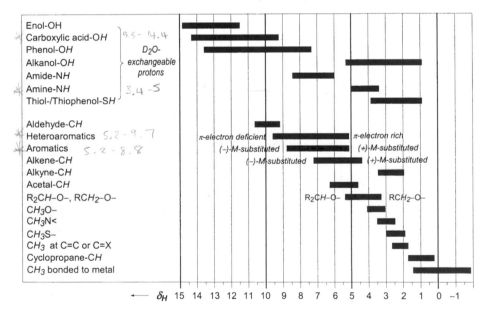

Small shift values for CH or CH_2 protons may indicate cyclopropane units. Proton shifts distinguish between alkyne CH (generally $\delta_H = 2.5 - 3.2$), alkene CH (generally $\delta_H = 4.5 - 6$) and aromatic/heteroaromatic CH ($\delta_H = 6 - 9.5$), and also between π-electron-rich (pyrrole, furan, thiophene, $\delta_H = 6 - 7$) and π-electron-deficient heteroaromatic compounds (pyridine, $\delta_H = 7.5 - 9.5$).

2.1.2 Deuterium exchange

Protons which are bonded to heteroatoms (XH protons, $X = O, N, S$) can be identified in the ^1H NMR spectrum by using deuterium exchange (treatment of the sample with a small amount of D_2O or CD_3OD). After the deuterium exchange:

$$R-XH \; + \; D_2O \; \rightleftharpoons \; R-XD \; + \; HDO$$

the XH proton signals in the ^1H NMR spectrum disappear. Instead, the HDO signal appears at approximately $\delta_H = 4.8$. Those protons which can be identified by D_2O exchange are indicated as such in Table 2.1. As a result of D_2O exchange, XH protons are often not detected in the ^1H NMR spectrum if this is obtained using a deuteriated protic solvent (e.g. CD_3OD).

2.1.3 ^{13}C Chemical shifts

The ^{13}C chemical shift ranges for organic compounds [4-6] in Table 2.2 show that many carbon-containing functional groups can be identified by the characteristic shift values in the ^{13}C NMR spectra.

For example, various carbonyl compounds have distinctive shifts. Ketonic carbonyl functions appear as singlets falling between $\delta_C = 190$ and 220, with cyclopentanone showing the largest shift; although aldehyde signals between $\delta_C = 185$ and 205 overlap with the shift range of keto carbonyls, they appear in the coupled ^{13}C NMR spectrum as doublet CH signals. Quinone carbonyls occurs between $\delta_C = 180$ and 190 while the carboxy C atoms of carboxylic acids and their derivatives are generally found between $\delta_C = 160$ and 180. However, the ^{13}C signals of phenoxy carbon atoms, carbonates, ureas (carbonic acid derivatives), oximes and other imines also lie at about $\delta_C = 160$ so that additional information such as the empirical formula may be helpful for structure elucidation.

Other functional groups that are easily differentiated are cyanide ($\delta_C = 110\text{-}120$) from isocyanide ($\delta_C = 135\text{-}150$), thiocyanate ($\delta_C = 110\text{-}120$) from isothiocyanate ($\delta_C = 125\text{-}140$), cyanate ($\delta_C = 105\text{-}120$) from isocyanate ($\delta_C = 120\text{-}135$) and aliphatic C atoms which are bonded to different heteroatoms or substituents (Table 2.2). Thus ether-methoxy generally appears between $\delta_C = 55$ and 62, ester-methoxy at $\delta_C = 52$; N-methyl generally lies between $\delta_C = 30$ and 45 and S-methyl at about $\delta_C = 25$. However, methyl signals at $\delta_C = 20$ may also arise from methyl groups attached to C=X or C=C double bonds, e.g. as in acetyl, CH_3–CO–.

If an H atom in an alkane R–H is replaced by a substituent X, the ^{13}C chemical shift δ_C in the α-position increases proportionally to the electronegativity of X ($-I$ effect). In the β-position, δ_C generally also increases, whereas it decreases at the C atom γ to the substituent (γ-effect, see Section 2.3.4). More remote carbon atoms remain almost uninfluenced ($\Delta\delta_C \sim 0$).

In contrast to ^1H shifts, ^{13}C shifts cannot in general be used to distinguish between aromatic and heteroaromatic compounds on the one hand and alkenes on the other (Table 2.2). Cyclopropane carbon atoms stand out, however, by showing particularly small shifts in both the ^{13}C and the ^1H NMR spectra. By analogy with their proton resonances, the ^{13}C chemical shifts of π electron-deficient heteroaromatics (pyridine type) are larger than those of π electron-rich heteroaromatic rings (pyrrole type).

Table 2.2. ^{13}C chemical shift ranges for organic compounds

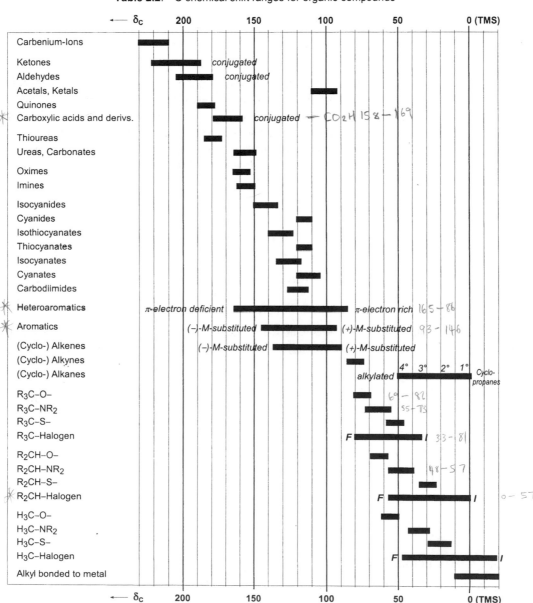

R₃CH occurs at 30-50.

Substituent effects (substituent increments) tabulated in more detail in the literature [1-6] demonstrate that ^{13}C chemical shifts of individual carbon nuclei in alkenes and aromatic as well as heteroaromatic compounds can be predicted approximately by means of mesomeric effects (resonance effects). Thus, an electron donor substituent D [D = OCH_3, SCH_3, $N(CH_3)_2$] attached to a C=C double bond shields the β-C atom and the β-*proton* (+M effect, smaller shift), whereas the α-position is deshielded (larger shift) as a result of substituent electronegativity (–I effect).

Donor in α shields in β postion    Acceptor in α deshields in β postion

The reversed polarity of the double bond is induced by a π electron-accepting substituent A (A = C=O, C=N, NO_2): the carbon and proton in the β-position are deshielded (–M effect, larger shifts).

These substituents have analogous effects on the C atoms of aromatic and heteroaromatic rings. An electron donor D (see above) attached to the benzene ring deshields the (substituted) α-C atom (–I effect). In contrast, in the *ortho* and *para* positions (or comparable positions in heteroaromatic rings) it causes a shielding (+M effect, smaller ^1H and ^{13}C shifts), whereas the *meta* positions remain almost unaffected.

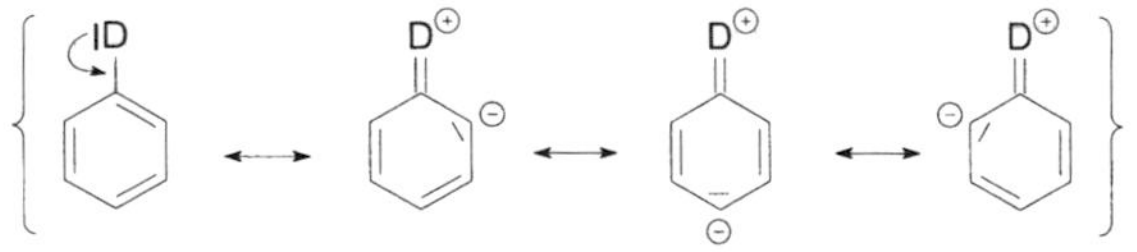

(+)-M-substituent (electron donor D) bonded to the benzene ring:
$\delta_H < 7.26$; $\delta_C < 128.5$

An electron-accepting substituent A (see above) induces the reverse deshielding in *ortho* and *para* positions (–M effect, larger ^1H and ^{13}C shifts), again with no significant effect on *meta* positions.

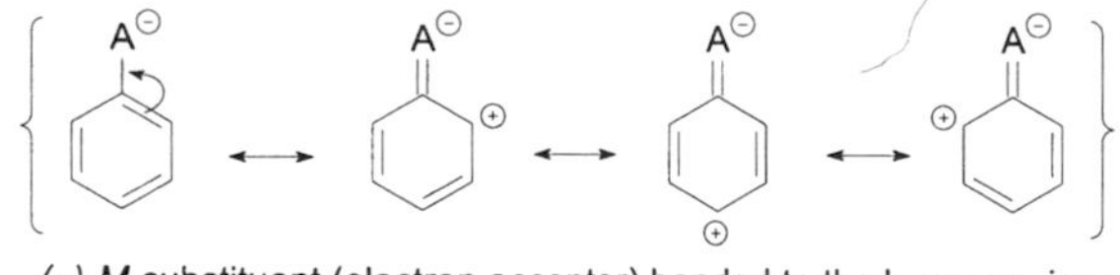

(–)-M-substituent (electron acceptor) bonded to the benzene ring:
$\delta_H > 7.26$; $\delta_C > 128.5$

2.1.4 ^{15}N Chemical shifts

Frequently the ^{15}N chemical shifts [7-9] (Table 2.3) of molecular fragments and functional groups containing nitrogen complement their ^1H and ^{13}C shifts. The ammonia scale [7] of ^{15}N shifts used in

Table 2.3 shows very obvious parallels with the TMS scale of ^{13}C shifts. Thus, the ^{15}N shifts (Table 2.3) decrease in size in the sequence nitroso, nitro, imino, amino, following the corresponding behaviour of the ^{13}C shifts of carbonyl, carboxy, alkenyl and alkyl carbon atoms (Table 2.2).

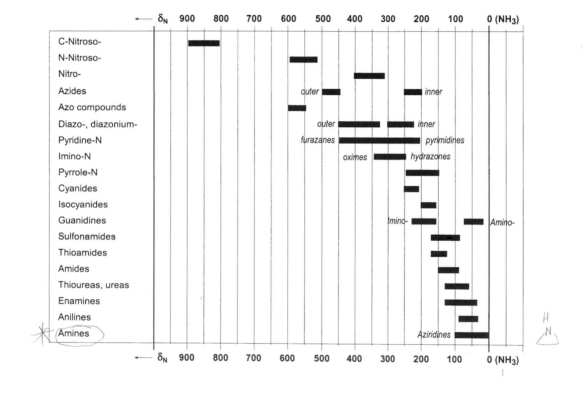

Table 2.3. ^{15}N Chemical shift ranges for organonitrogen compounds

The decrease in shifts found in ^{13}C NMR spectra in the sequence

$$\delta_{\text{alkenes, aromatics}} > \delta_{\text{alkynes}} > \delta_{\text{alkanes}} > \delta_{\text{cyclopropanes}}$$

also applies to the analogous *N*-containing functional groups, ring systems and partial structures (Tables 2.2 and 2.3):

$$\delta_{\text{imines, pyridines}} > \delta_{\text{nitriles}} > \delta_{\text{amines}} > \delta_{\text{aziridines}}$$

2.2   Skeletal structure (atom connectivities)

2.2.1   *HH* Multiplicities

The splitting (signal multiplicity) of ^1H resonances often reveals the spatial proximity of the protons involved. Thus it is possible to identify structural units such as those which often occur in organic molecules simply from the appearance of multiplet systems and by using the $(n+1)$ rule.

The simplest example is the *AX* or *AB* two-spin system for all substructures containing two protons two, three or four bonds apart from each other, according to *geminal*, *vicinal* or *w* coupling. Figure 2.1 shows the three typical examples: (**a**) the *AX* system, with a large shift difference ($v_X - v_A$) between the coupled protons H^A and H^X in relation to their coupling constant J_{AX}; (**b**) the *AB* system, with a smaller shift difference ($v_B - v_A$) of the coupled nuclei (H^A and H^B) relative to their coupling constant J_{AB}, and (**c**) the *AB* system, with a very small shift difference [$(v_B - v_A) \leq J_{AB}$] verging on the A_2 case, whereby the outer signals are very strongly suppressed by the roofing effect (*AB* effect). Figure 2.2 shows the ^1H NMR partial spectra of a few more structural units which can easily be identified.

Structure elucidation does not necessarily require the complete analysis of all multiplets in complicated spectra. If the coupling constants are known, the characteristic fine structure of the single multiplet almost always leads to identification of a molecular fragment and, in the case of alkenes and aromatic or heteroaromatic compounds, it may even lead to the elucidation of the complete substitution pattern.

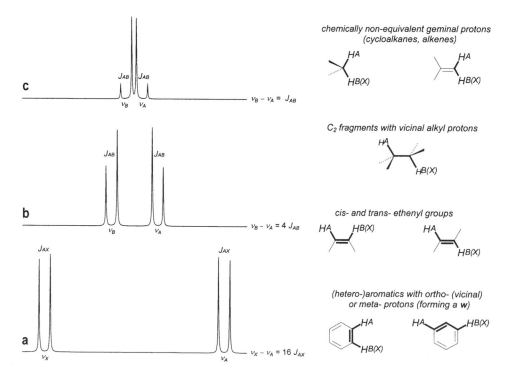

Figure 2.1. *AX (AB)* systems and typical molecular fragments

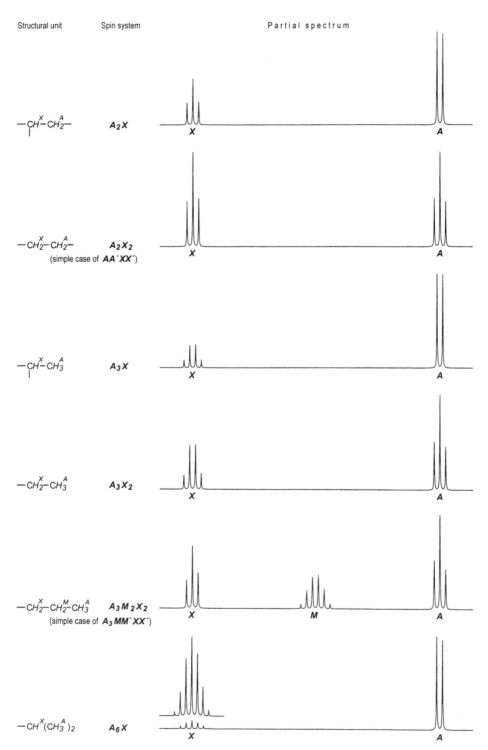

Figure 2.2. Easy to recognise A_mX_n systems and their typical molecular fragments

2.2.2   C*H* Multiplicities

The multiplicities of ^{13}C signals due to $^1J_{CH}$ coupling (splitting occurs due to C*H* coupling across one bond) indicates the bonding mode of the C atoms, whether quaternary (R$_4$C, singlet S), tertiary (R$_3$C*H*, doublet D), secondary (R$_2$C*H*$_2$, triplet T) or primary (RC*H*$_3$, quartet Q).

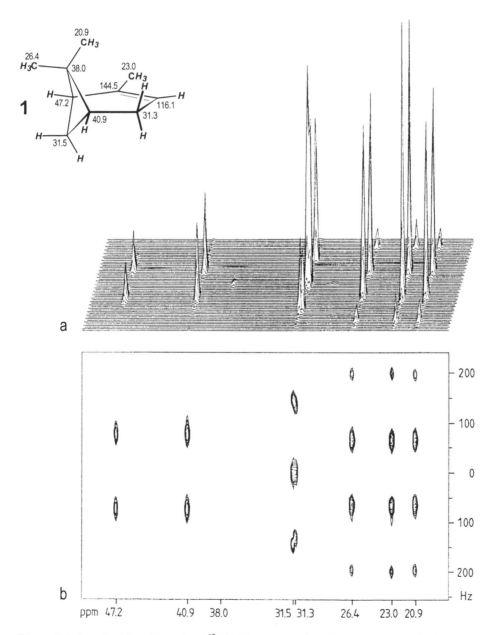

Figure 2.3. *J*-resolved two-dimensional ^{13}C NMR spectra series of α-pinene (**1**) [in (CD$_3$)$_2$CO, 25 °C, 50 MHz].
(**a**) Stacked plot; (**b**) contour plot

Coupled ^{13}C NMR spectra which have been enhanced by NOE are suitable for analysis of *CH* multiplicities (gated decoupling) [5,6]. Where the sequence of signals in the spectra is too dense, evaluation of spin multiplicities may be hampered by overlapping. In the past this has been avoided by compression of the multiplet signals using off-resonance decoupling [5,6] of the protons. More modern techniques include the *J*-modulated spin-echo technique (attached proton test, APT) [10,11] and *J*-resolved two-dimensional ^{13}C NMR spectroscopy [12,13], which both use *J*-modulation [14,15]. Figure 2.3, shows a series of *J*-resolved ^{13}C NMR spectra of α-pinene (**1**) as a contour plot and as a stacked plot. The purpose of the experiment is apparent; ^{13}C shift and J_{CH} coupling constants are shown in two frequency dimensions so that signal overlaps occur less often.

The *J*-modulated spin-echo [10,11] and the more frequently used DEPT experiment [14,15] are pulse sequences, which transform the information of the *CH* signal multiplicity and of spin-spin coupling into phase relationships (positive and negative amplitudes) of the ^{13}C signals in the proton-decoupled ^{13}C NMR spectra. The DEPT experiment benefits from a ^1H–^{13}C polarisation transfer which increases the sensitivity by up to a factor of 4. For this reason, this technique provides the quickest way of determining the $^{13}C/H$ multiplicities. Figure 2.4 illustrates the application of the DEPT technique to the analysis of the *CH* multiplets of α-pinene (**1**). Routinely the result will be the subspectrum (**b**) of all *CH* carbon atoms in addition to a further subspectrum (**c**), in which, besides the *CH* carbon atoms, the *CH₃* carbon atoms also show positive amplitude, whereas the CH_2 carbon atoms appear as negative. Quaternary C atoms do not appear in the DEPT subspectra; accordingly, they may be identified as the signals which appear additionally in the ^1H broadband decoupled ^{13}C NMR spectra (e.g. spectrum **a** in Fig. 2.4).

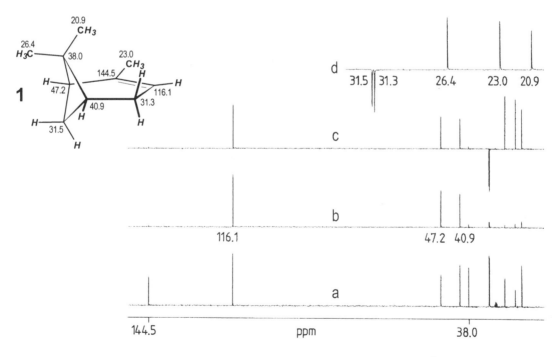

Figure 2.4. *CH* multiplicities of α-pinene (**1**) [hexadeuterioacetone, 25 °C, 50 MHz]. (**a**) ^1H broadband decoupled ^{13}C NMR spectrum; (**b**) DEPT subspectrum of *CH;* (**c**) DEPT subspectrum of all C atoms which are bonded to *H* (*CH* and *CH₃* positive, *CH₂* negative); (**d**) an expansion of a section of (**c**). Signals from two quaternary C atoms, three *CH* units, two *CH₂* units and three *CH₃* units can be seen

Figure 2.4 illustrates the usefulness of *CH* multiplicities for the purpose of structure elucidation. The addition of all C, *CH*, *CH$_2$* and *CH$_3$* units leads to a part formula C_xH_y,

$$2C + 3CH + 2CH_2 + 3CH_3 \; = \; C_2 + C_3H_3 + C_2H_4 + C_3H_9 = C_{10}H_{16}$$

which contains all of the *H* atoms which are bonded to C. Hence the result is the formula of the hydrocarbon part of the molecule, e.g. that of α-pinene (**1**, Fig. 2.4).

If the *CH* balance given by the *CH* multiplicities differs from the number of *H* atoms in the molecular formula, then the additional *H* atoms are bonded to heteroatoms. The ^{13}C NMR spectra in Fig. 2.5 show, for example, for isopinocampheol (**2**), $C_{10}H_{18}O$, a quaternary C atom (C), four *CH* units (C_4H_4), two *CH$_2$* units (C_2H_4) and three *CH$_3$* groups (C_3H_9). In the *CH* balance, $C_{10}H_{17}$, one *H* is missing when compared with the molecular formula, $C_{10}H_{18}O$; to conclude, the compound contains one O*H* group.

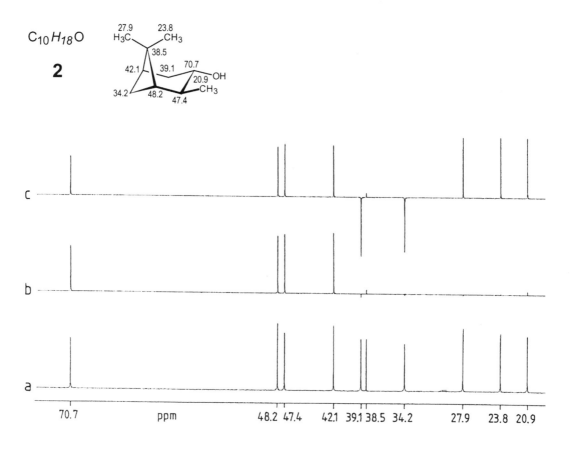

Figure 2.5. *CH* multiplicities of isopinocampheol (**2**), $C_{10}H_{18}O$ [(CD$_3$)$_2$CO, 25 °C, 50 MHz]. (**a**) ^1H broadband decoupled ^{13}C NMR spectrum; (**b**) DEPT *CH* subspectrum; (**c**) DEPT subspectrum of all C atoms which are bonded to *H* (*CH* and *CH$_3$* positive, *CH$_2$* negative)

2.2.3   HH Coupling constants

Since spin-spin coupling [2,3] through bonds occurs because of the interaction between the magnetic moments of the atomic nuclei and the bonding electrons, the coupling constants [2,3] reflect the bonding environments of the coupled nucei. In ^1H NMR spectroscopy *geminal* coupling through *two* bonds ($^2J_{HH}$) and *vicinal* coupling through *three* bonds ($^3J_{HH}$) provide insight into the nature of these bonds.

Table 2.4. Typical *HH* coupling constants (*Hz*) of some units in alicycles, alkenes and alkynes [2,3]

Geminal HH coupling, $^2J_{HH}$, depends characteristically on the polarity and hybridisation of the C atom on the coupling path and also on the substituents and on the *HCH* bond angle. Thus $^2J_{HH}$ coupling can be used to differentiate between a cyclohexane (*–12.5 Hz*), a cyclopropane (*–4.5 Hz*) or an alkene (*2.5 Hz*), and to show whether electronegative heteroatoms are bonded to methylene groups (Table 2.4). In cyclohexane and norbornane derivatives the *w*-shaped arrangement of the bonds between protons attached to alternate C atoms leads to distinctive $^4J_{HH}$ coupling (*w*-couplings, Table 2.4).

Vicinal HH coupling constants, $^3J_{HH}$, are especially useful in determining the *relative configuration* (see Section 2.3.1). However, they also reflect a number of other distinguishing characteristics, e.g. the ring size for cycloalkenes (a low value for small rings) and the α-position of electronegative heteroatoms in heterocycles which is reflected by remarkably small coupling constants $^3J_{HH}$ (Table 2.5).

The coupling constants of *ortho* ($^3J_{HH}$ = 7 Hz), *meta* ($^4J_{HH}$ = 1.5 Hz) and *para* protons ($^5J_{HH} \leq 1$ Hz) in benzene and naphthalene ring systems are especially useful in structure elucidation (Table 2.5). With naphthalene and other condensed (hetero-) aromatics, a knowledge of *'zig zag'* coupling ($^5J_{HH}$ = 0.8 Hz) is helpful in deducing substitution patterns.

Table 2.5. Typical *HH* coupling constants (*Hz*) of aromatic and heteroaromatic compounds [2,3]

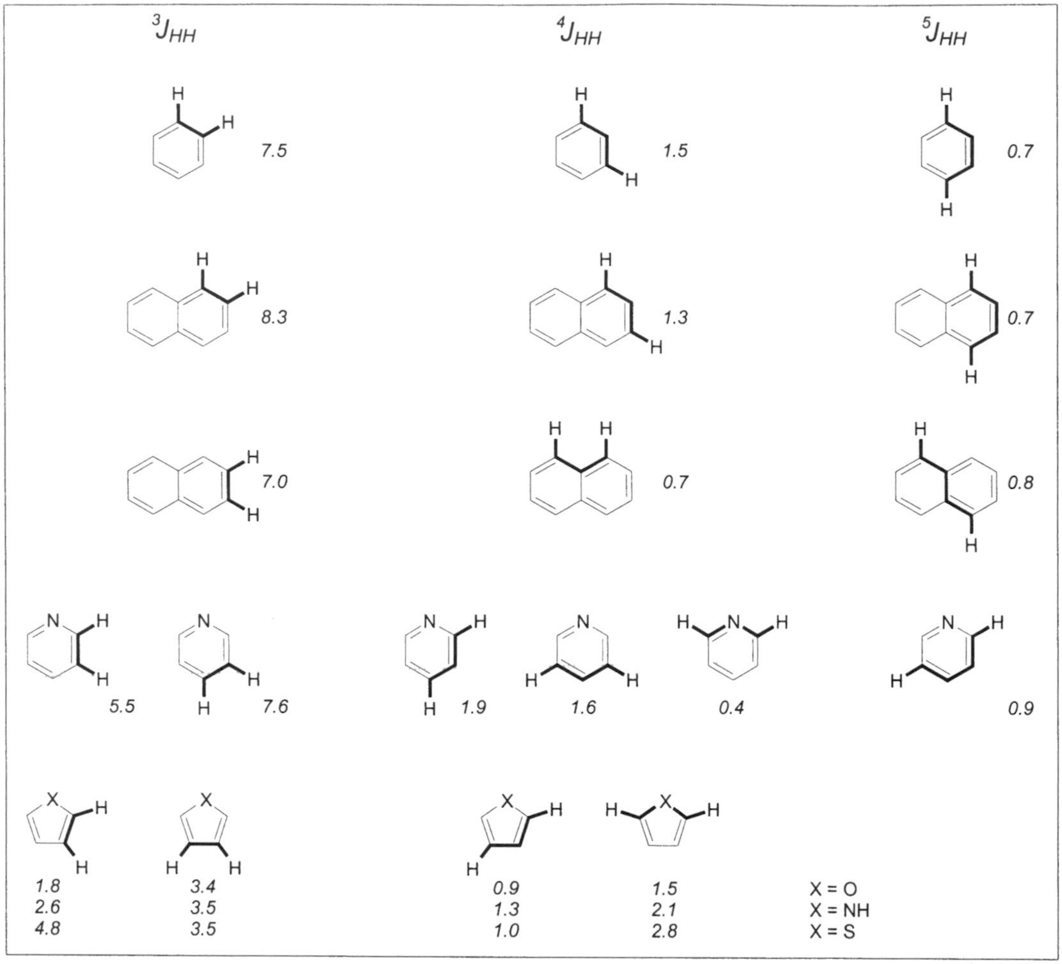

$^3J_{HH}$	$^4J_{HH}$	$^5J_{HH}$
(benzene) 7.5	(benzene) 1.5	(benzene) 0.7
(naphthalene) 8.3	(naphthalene) 1.3	(naphthalene) 0.7
(naphthalene) 7.0	(naphthalene) 0.7	(naphthalene) 0.8
(pyridine) 5.5 (pyridine) 7.6	(pyridine) 1.9 (pyridine) 1.6 (pyridine) 0.4	(pyridine) 0.9
1.8 / 2.6 / 4.8 3.4 / 3.5 / 3.5	0.9 / 1.3 / 1.0 1.5 / 2.1 / 2.8	X = O / X = NH / X = S

The *HH* coupling constants of pyridine (Table 2.5) reflect the positions of the coupling protons relative to the nitrogen ring. There is a particularly clear difference here between the protons in the 2- and 3-positions ($^3J_{HH}$ = 5.5 Hz) and those in the 3- and 4-positions ($^3J_{HH}$ = 7.6 Hz). Similarly, *HH* coupling constants in five-membered heteroaromatic rings such as thiophene, pyrrole and furan can be distinguished because of the characteristic effects of the electronegative heteroatoms on their $^3J_{HH}$ couplings (Table 2.5); in particular the $^3J_{HH}$ coupling of the protons in the 2- and 3-positions, allow the heteroatoms to be identified (the more electronegative the heteroatom, the smaller is the value of $^3J_{HH}$).

In the case of alkenes and aromatic and heteroaromatic compounds, analysis of a single multiplet will often clarify the complete substitution pattern. A few examples will illustrate the procedure.

If, for example, four signals are found in regions appropriate for benzene ring protons (δ_H = 6 - 9, four protons on the basis of the height of the integrals), then the sample may be a disubstituted benzene (Fig. 2.6). The most effective approach is to analyse a multiplet with a clear fine structure and as many coupling constants as possible, e.g. consider the threefold doublet at $\delta_H = 7.5$ (Fig. 2.6 **a**); it shows two *ortho* couplings (*8.0* and *7.0 Hz*) and one *meta* coupling (*2.5 Hz*); hence relative to the *H* atom with a shift value of $\delta_H = 7.5$, there are two protons in *ortho* positions and one in a *meta* position; to conclude, the molecule is an *ortho*-disubstituted benzene (*o*-nitrophenol, **3**).

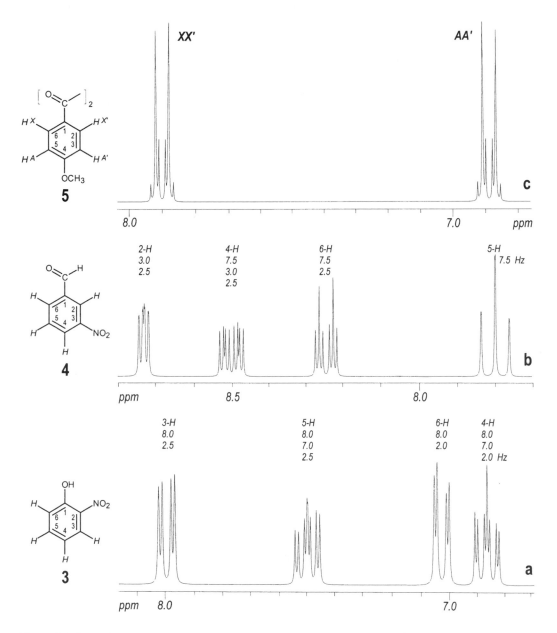

Figure 2.6. ^1H NMR spectra of disubstituted benzene rings [CDCl$_3$, 25 °C, 200 MHz]. (**a**) *o*-Nitrophenol (**3**); (**b**) *m*-nitrobenzaldehyde (**4**); (**c**) 4,4'-dimethoxybenzil (**5**)

A *meta* disubstituted benzene (Fig. 2.6 **b**) shows only two *ortho* couplings ($^3J_{HH} = 7.5\ Hz$) for one signal ($\delta_H = 7.8$) whereas another signal ($\delta_H = 8.74$) exhibits only two *meta* couplings ($^4J_{HH} = 3.0$ and *2.5 Hz*). In both cases one observes either a triplet ($\delta_H = 7.8$) or a doublet of doublets ($\delta_H = 8.74$) depending on whether the couplings ($^3J_{HH}$ or $^4J_{HH}$) are equal or different.

The *AA'XX'* systems (Section 1.5) [2,3] which are normally easily recognisable from their symmetry identify *para-disubstituted* benzenes such as 4,4'-dimethoxybenzil (**5**) or 4- substituted pyridines.

This method of focusing on a ^1H multiplet of clear fine structure and revealing as many *HH* coupling constants as possible affords the substitution pattern for an alkene or an aromatic or a hetero-aromatic compound quickly and conclusively. One further principle normally indicates the *geminal, vicinal* and *w* relationships of the protons of a molecule, the so called *HH connectivities*, i.e. that *coupled nuclei have identical coupling constants*. Accordingly, once the coupling constants of a multiplet have all been established, the appearance of one of these couplings in another multiplet identifies (and assigns) the coupling partner. This procedure, which also leads to the solutions to problems 1-12, may be illustrated by means of two typical examples.

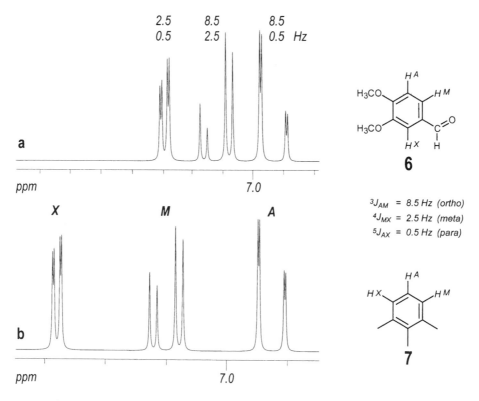

Figure 2.7. ^1H NMR spectrum of 3,4-dimethoxybenzaldehyde (**6**) [aromatic shift range, CDCl$_3$, 25 °C, (a) 100 MHz, (b) 200 MHz]

In Fig. 2.7 the ^1H signal with a typical aromatic proton shift of $\delta_H = 7.1$ shows a doublet of doublets with *J*-values of *8.5 Hz* (*ortho* coupling, $^3J_{HH}$) and *2.5 Hz* (*meta* coupling, $^4J_{HH}$). The ring proton in question therefore has two protons as coupling partners, one in the *ortho* position (*8.5*

Hz) and another in the *meta* position (*2.5 Hz*), and moreover these are in such an arrangement as to make a second *ortho* coupling impossible. Thus the benzene ring is 1,2,4-trisubstituted (**6**). The ring protons form an *AMX* system, and, in order to compare the change of frequency dispersion and 'roofing' effects with increasing magnetic field strength, this is shown first at 100 MHz and then also at 200 MHz. The *para* coupling $^5J_{AX'}$ which is less frequently visible, is also resolved. From the splitting of the signal at $\delta_H = 7.1$ (H^M) a 1,2,3-trisubstituted benzene ring (**7**) might have been considered. In this case, however, the *ortho* proton (H^A) would have shown a second *ortho* coupling to the third proton (H^X).

The application of the principle that coupled nuclei will have the same coupling constant enables the ^1H NMR spectrum to be assigned completely (Fig. 2.7). The *ortho* coupling, $^3J_{AM} = 8.5$ Hz, is repeated at $\delta_H = 6.93$ and allows the assignment of H^A; the *meta* coupling, $^4J_{MX} = 2.5$ Hz, which appears again at $\delta_H = 7.28$, gives the assignment of H^X.

The four signals in the ^1H NMR spectrum of a pyridine derivative (Fig. 2.8) show first that it is a 2- or 3-monosubstituted derivative; a 4-monosubstituted pyridine would display an *AA'XX'* system. The signal with the smallest shift ($\delta_H = 7.16$) splits into a threefold doublet with coupling constants *8.1*, *4.8* and *0.7 Hz* . The two $^3J_{HH}$ couplings of *8.1* and *4.8 Hz* unequivocally belong to a β proton of the pyridine ring according to Table 2.5. Step by step assignment of all three couplings (Fig. 2.8) leads to a pyridine ring **8** substituted in the 3-position. Again, signals are assigned following the principle that coupled nuclei will have the same coupling constant; the coupling constants identified from Table 2.5 for the proton at $\delta_H = 7.16$ are then sought in the other multiplets.

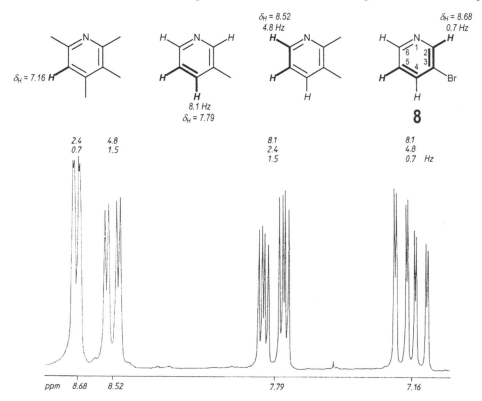

Figure 2.8. ^1H NMR spectrum of 3-bromopyridine (**8**) [CDCl$_3$, 25 °C , 90 MHz]

2.2.4   *CH* Coupling constants

One-bond *CH* coupling constants J_{CH} ($^1J_{CH}$) are proportional to the *s* character of the hybrid bonding orbitals of the coupling carbon atom, (Table 2.6, from left to right) [4-6,16], according to

$$J_{CH} = 500\ s \qquad\qquad (1)$$

where $s = 0.25$, 0.33 and 0.5 for sp^3-, sp^2- and sp-hybridised C atoms, respectively.

With the help of these facts, it is possible to distinguish between alkyl-C ($J_{CH} \sim 125$ Hz), alkenyl- and aryl-C ($J_{CH} \sim 165$ Hz) and alkynyl-C ($J_{CH} \sim 250$ Hz), e.g. as in problem 15.

It is also useful for structure elucidation that J_{CH} increases with the electronegativity of the heteroatom or substituent bound to the coupled carbon atom (Table 2.6, from top to bottom).

Table 2.6. Structural features (carbon hybridisation, electronegativity, ring size) and typical one-bond *CH* coupling constants J_{CH} (Hz) [4-6,16]

From typical values for J_{CH} coupling, Table 2.6 shows:

In the chemical shift range for aliphatic compounds
cyclopropane rings (*ca* 160 Hz reflect large *s* character of bonding hybrid orbitals);
oxirane (epoxide) rings (*ca* 175 Hz additionally reflect electronegativity of ring oxygen atom);
cyclobutane rings (*ca* 135 Hz);
O-alkyl groups (145-150 Hz);
N-alkyl groups (140 Hz);
acetal-C atoms (*ca* 170 Hz at $\delta_C = 100$);
terminal ethynyl groups (*ca* 250 Hz).

In the chemical shift range for alkenes and aromatic and heteroaromatic compounds

enol ether fragments (furan, pyrone, isoflavone, 195-200 Hz);

2-unsubstituted pyridine and pyrrole (*ca* 180 Hz);

2-unsubstituted imidazole and pyrimidine (> 200 Hz).

Geminal CH coupling $^2J_{CH}$ becomes more positive with increasing CC*H* bond angle and with decreasing electronegativity of the substituent on the coupling C. This property enables a distinction to be made inter alia between the substituents on the benzene ring or between heteroatoms in five-ring heteroaromatics (Table 2.7). From Table 2.7, those $^2J_{CH}$ couplings which may be especially clearly distinguished and diagnostic are:

β-C atoms in imines (e.g. C-3 in pyridine: 7 Hz);

α-C atoms in aldehydes (25 Hz);

substituted (non-protonated) C atoms of terminal ethynyl groups (40-50 Hz).

Table 2.7. Structural features and *geminal* (two-bond) C*H* coupling constants $^2J_{CH}$ (Hz) [4-6,16]

Vicinal CH couplings $^3J_{CH}$ depend not only on the configuration of the coupling C and *H* (Table 2.8; see Section 2.3.2), but also on the nature and position of substituents: an electronegative substituent raises the $^3J_{CH}$ coupling constant on the coupled C and lowers it on the coupling path, e.g. in alkenes and benzene rings (Table 2.8). An imino-*N* on the coupling path (e.g. from C-2 to 6-*H* in pyridine, Table 2.8) is distinguished by a particularly large $^3J_{CH}$ coupling constant (12 Hz).

In the ^{13}C NMR spectra of benzene derivatives, apart from the $^1J_{CH}$, only the *meta* coupling ($^3J_{CH}$, but not $^2J_{CH}$) is usually resolved. A benzenoid C*H,* from whose perspective the *meta* positions are substituted, usually appears as a $^1J_{CH}$ doublet without additional splitting, e.g. in the case of 3,4-dimethoxy-β-methyl-β-nitrostyrene (**9**, Fig. 2.9) the carbon atom C-5 generates a doublet at δ$_C$ = 111.5 in contrast to C-2 at δ$_C$ = 113.5 which additionally splits into a triplet. The use of C*H* coupling constants as criteria for assigning a resonance to a specific position is illustrated by this example.

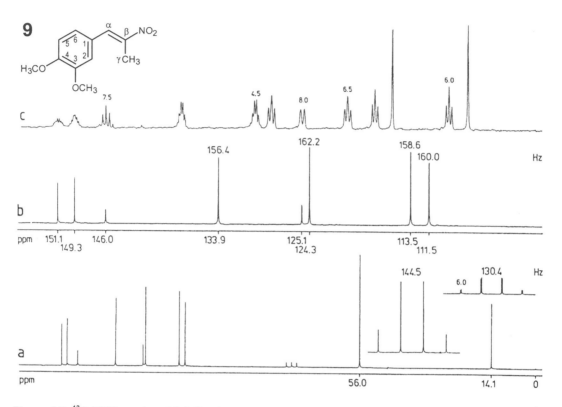

Figure 2.9. ^{13}C NMR spectra of 3,4-dimethoxy-β-methyl-β-nitrostyrene (**9**) [CDCl$_3$, 25 °C, 20 MHz]. (**a, b**) ^1H broadband decoupled, (**a**) complete spectrum with CH$_3$ quartets at δ$_C$ = 14.1 and 56.0; (**b, c**) decoupled and coupled partial spectrum of benzenoid and alkene carbon atoms, (**c**) obtained by 'gated' decoupling

Assignments:

C	δ$_C$ (ppm)	Multiplet	J_{CH} (Hz)	Multiplet	$^{3(2)}J_{CH}$ (Hz)	(coupling protons)
C-1	125.1	S		d	8.0	(5-H)
C-2	113.5	D	158.6	't' [a]	6.0	(6-H, α-H)
C-3	149.3	S		m		(5-H, 3-OCH$_3$)
C-4	151.1	S		m		(2-H, 6-H, 4-OCH$_3$)
C-5	111.5	D	160.0			
C-6	124.3	D	162.4	't' [a]	6.5	(2-H, α-H)
C-α	133.9	D	156.4	'sxt' [a]	4.5	(2-H, 6-H, β-CH$_3$)
C-β	146.0	S		'qui' [a]	7.5	(α-H, β-CH$_3$)
C-γ	14.1	Q	130.4	d	6.0	(α-H)
(OCH$_3$)$_2$	56.0	Q	144.5			

[a] The quotation marks indicate that the coupling constants are virtually the same for non-equivalent protons. C-β should, for example, split into a doublet ($^2J_{CH}$ to α-H) of quartets ($^2J_{CH}$ to β-CH$_3$); since both couplings have the same value (7.5 Hz), a pseudoquintet 'qui' is observed.

Table 2.8. Structural features and *vicinal* (three-bond) C*H* coupling constants $^3J_{CH}$ (Hz) [4-6,16]

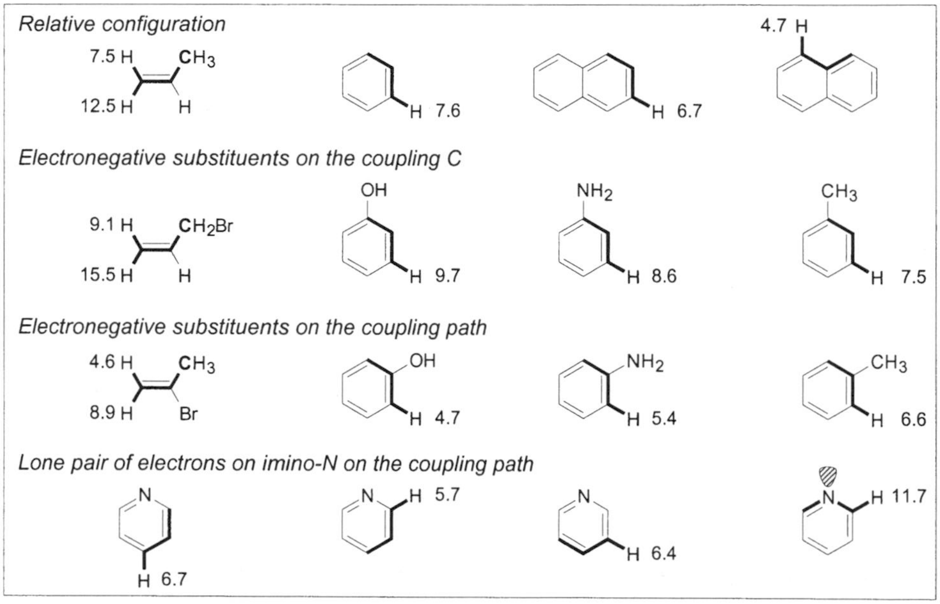

Usually there is no splitting between two exchangeable X*H* protons (X = O, N, S) and C atoms through two or three bonds ($^2J_{CH}$ or $^3J_{CH}$), unless solvation or an intramolecular *H* bridge fix the X*H* proton in the molecule. Thus the C atoms *ortho* to the hydroxy group show $^3J_{CH}$ coupling to the hydrogen bonding O*H* proton in salicylaldehyde (**10**), whose values reflect the relative configurations of the coupling partners. This method may be used, for example, to identify and assign the resonances in problem 17.

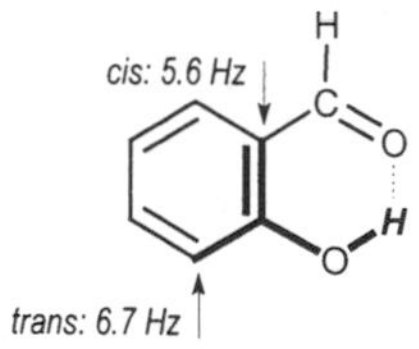

2.2.5   N*H* Coupling constants

Compared with ^1H and ^{13}C, the magnetic moment of ^{15}N is very small and has a negative value. The N*H* coupling constants are correspondingly smaller and their signs are usually the reverse of comparable *HH* and *CH* couplings [7-9]. Table 2.9 shows that the one-bond N*H* coupling, J_{NH}, is proportional to the *s*-character of the hybrid bonding orbital on *N* so a distinction can be made between amino- and imino-N*H*. Formamides can be identified by large $^2J_{NH}$ couplings between ^{15}N and the formyl proton. The $^2J_{NH}$ and $^3J_{NH}$ couplings of pyrrole and pyridine are especially distinctive and reflect the orientation of the non-bonding electron pair on nitrogen (pyrrole: perpendicular to the ring plane; pyridine: in the ring plane; Table 2.9), a fact which can be exploited in the identification of heterocyclic compounds (problems 30 and 31).

Table 2.9. Structural features and typical N*H* coupling constants (Hz) [1]

2.2.6   *HH* COSY (*geminal, vicinal, w*-relationships of protons)

The *HH* COSY experiment [12-13,17-19] in proton magnetic resonance is a quick alternative to spin decoupling [2,3] in structure elucidation. 'COSY' is the acronym derived from **c**orrelation **s**pectroscopy. *HH* COSY correlates the ^1H shifts of the coupling protons of a molecule. The proton shifts are plotted on both frequency axes in the two-dimensional experiment. The result is a diagram with square symmetry (Fig. 2.10). The projection of the *one-dimensional* ^1H NMR spectrum appears on the diagonal (*diagonal signals*). In addition there are *correlation* or *cross signals* (off-diagonal signals) where the protons are coupled with one another. Thus the *HH* COSY diagram indicates *HH* connectivities, that is, *geminal, vicinal* and *w*-relationships of the *H* atoms of a molecule and the associated structural units.

An *HH* COSY diagram can be shown in perspective as a stacked plot (Fig. 2.10**a**). Interpretation of this neat, three-dimensional representation, where the signal intensity gives the third dimension, can prove difficult because of distortions in the perspective. The contour plot can be interpreted more easily. This shows the signal intensity at various cross-sections (contour plots, Fig. 2.10**b**). However the choice of the plane of the cross-section affects the information provided by an *HH* COSY diagram; if the plane of the cross-section is too high then the cross signals which are weak are lost; if it is too low, then weaker artefacts may be mistaken for cross signals.

Every *HH* coupling interaction can be identified in the *HH* COSY contour plot by two diagonal signals and the two cross signals of the coupling partners, which form the four corners of a square. The coupling partner (cross signal) of a particular proton generates a signal on the vertical or horizontal line from the relevant ^1H signal. In Fig. 2.10**b**, for example, the protons at δ_H =7.90 and 7.16 are found as coupling partners on both the vertical and the horizontal lines from the proton 2-*H* of quinoline (**11**) at δ_H = 8.76. Since 2-*H* (δ_H = 8.76) and 3-*H* (δ_H = 7.16) of the pyridine ring in **11** can be identified by the common coupling $^3J_{HH}$ = 5.5 Hz (Table 2.5), the *HH* relationship which is likewise derived from the *HH* COSY diagram confirms the location of the pyridine protons in **11a**. Proton 4-*H* of quinoline (δ_H = 7.90) shows an additional cross signal at δ_H = 8.03 (Fig. 2.10). If it is known that this so-called *zig-zag* coupling is attributable to the benzene ring proton 8-*H*

(**11b**), then two further cross signals from $\delta_H = 8.03$ (at $\delta_H = 7.55$ and 7.35) locate the remaining protons of quinoline (**11c**).

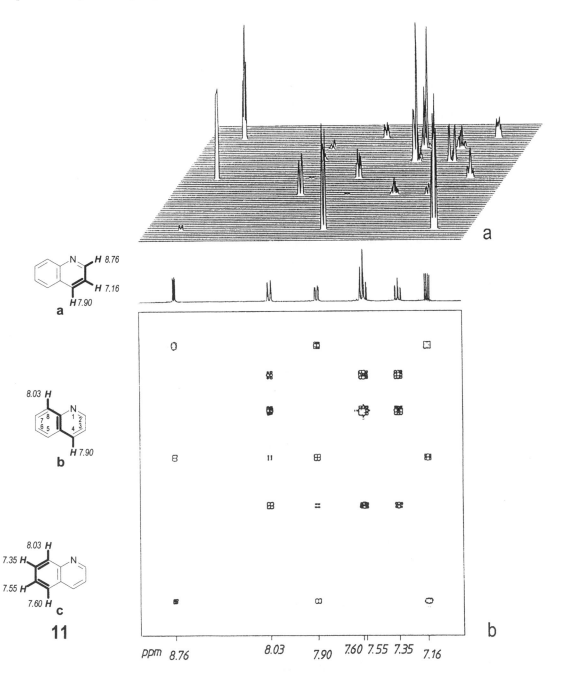

Figure 2.10. *HH* COSY diagram of quinoline (**11**) [(CD$_3$)$_2$CO, 95% v/v, 25°C, 400 MHz, 8 scans, 256 experiments]. (**a**) Stacked plot; (**b**) contour plot

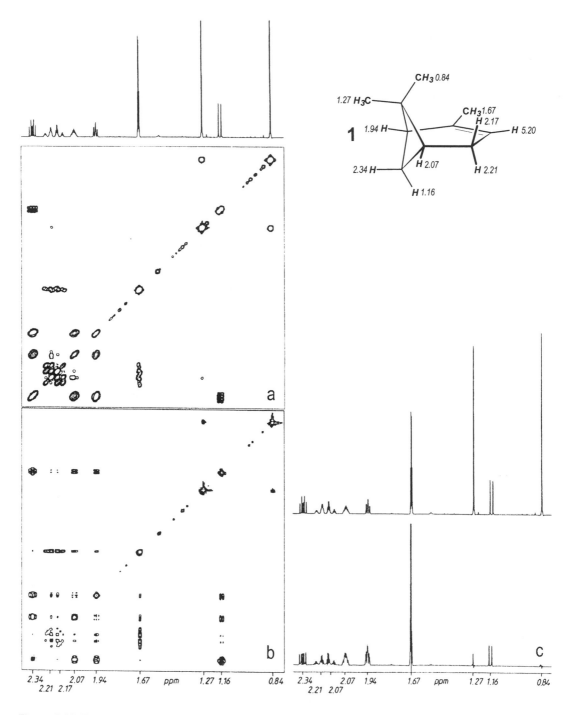

Figure 2.11. Proton-Proton shift correlations of α-pinene (**1**) [purity 99 %, CDCl₃, 5 % v/v, 25 °C, 500 MHz, 8 scans, 256 experiments]. (**a**) *HH* COSY; (**b**) *HH* TOCSY; (**c**) selective one-dimensional *HH* TOCSY, soft pulse irradiation at δ_H = 5.20 (signal not shown), compared with the ^1H NMR spectrum on top; deviations of chemical shifts from those in other experiments (Fig. 2.14, 2.16) arise from solvent effects

This example (Fig. 2.10) also shows the limitations of the *HH* COSY experiment: first, evaluation, without taking known shifts and possible couplings into account, is not always conclusive because the cross-sectional area of the cross signals may not reveal which specific couplings are involved; second, overlapping signals (e.g. $\delta_H = 7.55$ and *7.60* in Fig. 2.10) are not separated by *HH* COSY if the relevant protons couple to one another. If there is sufficient resolution, however, the fine structure of the multiplets may be recognised by the shapes of the diagonal and cross signals, e.g. in Fig. 2.10, at $\delta_H = 7.55$ there is a triplet, therefore the resonance at $\delta_H = 7.60$ is a doublet (see the shape of the signal on the diagonal at $\delta_H = 7.55$-*7.60* in Fig. 2.10).

In the case of (m+1)- and (n+1)-fold splittings of A_mX_n systems in one-dimensional ^1H NMR spectra the *HH* COSY plot gives, depending on the resolution, up to (m+1)(n+1)-fold splittings of the cross signals. If several small coupling constants contribute to a multiplet, the intensity of the cross signals in the *HH* COSY may be distributed into many multiplet signals so that even at a low-lying cross-section no cross signals appear in the contour diagram. Despite these limitations, structural fragments may almost always be derived by means of the *HH* COSY, so that with complementary information from other NMR experiments it is possible to deduce the complete structure. Thus the *HH* COSY is especially useful for solving the problems 12, 32, 33, 38, 46 and 52.

Modifications of the basic *HH* COSY [17] include "COSY 45" with an initial 45° pulse to suppress diagonal signals for a better separation of cross signals near the diagonal or "COSY with delay" to emphasise connectivities which are the result of smaller couplings. The "TOCSY" experiment (from total correlation spectroscopy) [2] detects cross signals of an individual proton with all other protons coupled to each other within a larger structural unit, e.g. a ring, the range depending on the adjustable time of isotropic magnetisation transfer. Thus, the *endo*-methylene proton of α-pinene at $\delta_H = 1.16$ displays only one cross signal with the *exo*-methylene proton ($\delta_H = 2.34$) due to *geminal* coupling in the *HH* COSY experiment (Fig. 2.11**a**), whereas additional cross signals with $\delta_H = 1.94, 2.07, 2.17$ and *2.21* appear in the *HH* TOCSY plot (Fig. 2.11**b**), thus revealing a considerably larger partial structure. One-dimensional *HH* TOCSY experiments involve irradiation with a selective ("soft") pulse adjustable to any non-overlapping proton signal of the molecule. Fig. 2.11 **c**, for example, detects all protons of the cyclohexene substructure including the methylene group in a one-dimensional *HH* TOCSY with selective irradiation of the alkene proton ($\delta_H = 5.20$).

2.2.7   CC INADEQUATE (CC bonds)

Once all of the CC bonds in a molecule are known, then its carbon skeleton is established. One way to identify the CC bonds would be to measure ^{13}C-^{13}C coupling constants, since these are the same for C atoms which are bonded to one another: identical coupling constants are known to identify the coupling partners (see Section 2.2.3). Unfortunately, this method is complicated by two factors: first, ^{13}C-^{13}C couplings, especially those in the aliphatic region, are nearly all the same ($J_{CC} = 35$-40 Hz [16], Fig. 2.12), provided that none of the coupling C atoms carries an electronegative substituent. Second, the occurrence of ^{13}C-^{13}C coupling requires the two nuclei to be directly bonded. However, given the low natural abundance of ^{13}C (1.1 % or 10^{-2}), the probability of a ^{13}C-^{13}C bond is only 10^{-4}. Splitting as a result of ^{13}C-^{13}C coupling therefore appears only as a weak feature in the spectrum (0.5 % intensity), usually in satellites which are concealed by noise at a distance of half the ^{13}C-^{13}C coupling constant on either side of the ^{13}C-^{12}C main signal (99 % intensity).

The one-dimensional variations of the INADEQUATE experiment [12,13,17,20] suppress the intense ^{13}C-^{12}C main signal, so that both AX and AB systems appear for all ^{13}C-^{13}C bonds in one spectrum. The two-dimensional methods [12,13,17,21,22] segregate these AB systems on the basis of their individual double quantum frequencies (the sum of the ^{13}C shifts of A and B) as a second dimension. Using the simple example of 1-butanol (**12**), Fig. 2.12**a** demonstrates the use of the two-dimensional INADEQUATE experiment for the purpose of structure elucidation. For every C–C bond the contour diagram gives an AB system parallel to the abscissa with double quantum frequency as ordinate. By following the arrows in Fig. 2.12**a**, the carbon-carbon bonds of 1-butanol can be derived immediately. The individual AB systems may also be plotted one-dimensionally (Fig. 2.12**b**); the ^{13}C-^{13}C coupling constants often provide useful additional information.

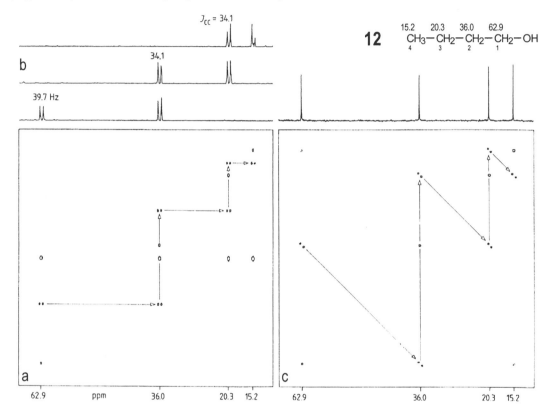

Figure 2.12. Two-dimensional (2D-)INADEQUATE diagram of 1-butanol (**12**) [(CD$_3$)$_2$CO, 95% v/v, 25°C, 50 MHz, 128 scans, 128 experiments]. (**a**) Contour plot with the AB systems of bonded C atoms on the horizontal axis; (**b**) plots of the three AB systems of the molecule obtained from (**a**); (**c**) contour plot of the symmetrised INADEQUATE experiment showing the AB or AX systems of bonded C atoms in the HH COSY format (cross signals on the axes perpendicular to the diagonal)

A variation on the INADEQUATE technique, referred to as symmetrised 2D INADEQUATE [17,21,22], provides a representation in the HH COSY format with its quadratic symmetry of the diagonal and cross signals. Here the one-dimensional ^1H broadband decoupled ^{13}C NMR spectrum is projected on to the diagonal and the AB systems of all C–C bonds of the molecule are projected on to individual orthogonals (Fig. 2.12**c**). Every C–C bond then gives a square defined by diagonal signals and off-diagonal AB patterns, and it is possible to evaluate as described for HH COSY.

A disadvantage is the naturally low sensitivity of the INADEQUATE technique. However, if one has enough substance (more than 5 mg per C atom, samples from syntheses), then the sophisticated experiment is justified as the solutions to the problems 21, 22, 35, 36 and 37 illustrate. Fig. 2.13 is intended to demonstrate the potential of this technique for tracing out a bicyclic carbon skeleton using the example of isopinocampheol (**2**). The evaluation of all CC-*AB* systems on the orthogonals leads to the eleven C–C bonds a-k. If all the C–C bonds which have been found are combined, then the result is the bicyclic system (a-h) and the three methyl substituents (i-k) of the molecule **2**.

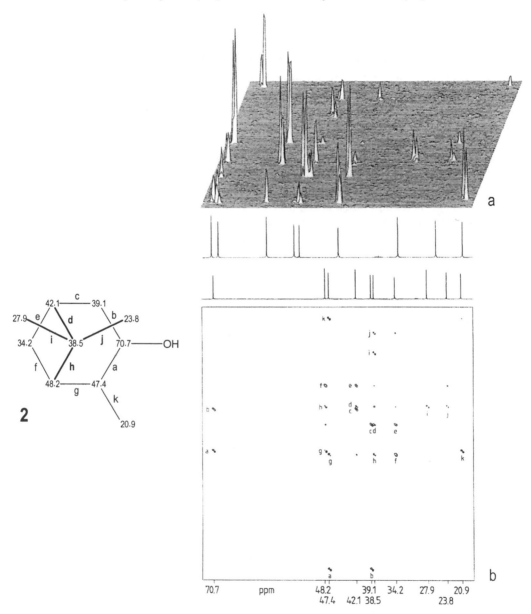

Figure 2.13. Symmetrised two-dimensional INADEQUATE experiment with isopinocampheol (**2**) [(CD₃)₂CO, 250 mg in 0.3 ml, 25 °C, 50 MHz, 256 scans and exp.]. (**a**) Stacked plot of the section between δ_C = 20.9 and 48.2; (**b**) complete contour plot with cross signal pairs labelled a-k for the 11 CC bonds of the molecule to facilitate the assignments sketched in formula **2**

The point of attachment of the *OH* group within the molecule (at $\delta_C = 70.7$) is revealed by the DEPT technique in Fig. 2.5. Figure 2.13 also shows the *AB* effect on the ^{13}C signals of neighbouring C atoms with a small shift difference (bond g with $\delta_C = 47.4$ and 48.2): the intense inner signals appear very clearly; the weak outer signals of the *AB* system of these two C atoms are barely recognisable except as dots. Additional cross signals without resolved doublet structure, e.g. between $\delta_C = 48.2$ and 42.1, are the result of smaller $^2J_{CC}$ and $^3J_{CC}$ couplings.

2.2.8   Two-dimensional carbon-proton shift correlation *via* one-bond C*H* coupling

The C*H* COSY technique [12,13,17,23] with carbon-13 detection and proton decoupling, and more sensitive and thus time saving methods with ^1H detection and ^{13}C decoupling denoted as *inverse* C*H* COSY, correlate ^{13}C shifts in one dimension with the ^1H shifts in the other *via* one-bond C*H* coupling J_{CH}.

The pulse sequence which is used to record C*H* COSY involves the ^1H-^{13}C polarisation transfer which is the basis of the DEPT sequence and which increases the sensitivity by a factor of up to four. Consequently, a C*H* COSY experiment does not require any more sample than a ^1H broadband decoupled ^{13}C NMR spectrum. The result is a two-dimensional C*H* correlation, in which the ^{13}C shift is mapped on to the abscissa and the ^1H shift is mapped on to the ordinate (or vice versa). The ^{13}C and ^1H shifts of the ^1H and ^{13}C nuclei which are bonded to one another are read as coordinates of the cross signal as shown in the C*H* COSY stacked plot (Fig. 2.14**b**) and the associated contour plots of the α-pinene (Fig. 2.14**a** and **c**). To evaluate them, one need only read off the coordinates of the correlation signals. In Fig. 2.14**c**, for example, the protons with shifts $\delta_H = 1.16$ (proton *A*) and *2.34* (proton *B* of an *AB* system) are bonded to the C atom at $\delta_C = 31.5$. Formula **1** shows all of the C*H* connectivities (C*H* bonds) of α-pinene which can be read from Fig. 2.14.

*H*C HMQC (heteronuclear multiple quantum coherence) and *H*C HSQC (heteronuclear single quantum coherence) are the acronyms of the pulse sequences used for inverse carbon-proton shift correlations. These sensitive inverse experiments detect one-bond carbon-proton connectivities within some minutes instead of some hours as required for C*H* COSY as demonstrated by an *H*C HSQC experiment with α-pinene in Fig. 2.15.

Carbon-proton shift correlations are particularly attractive for structure elucidation because they allow the shifts of two nuclei (^1H and ^{13}C) to be measured in a single experiment and within a feasible time scale. They determine all C*H* bonds (the C*H* connectivities) of the molecule, and hence provide an answer to the problem as to which ^1H nuclei are bonded to which ^{13}C nuclei. The ^1H multiplets, which so frequently overlap in the ^1H domain, are almost always separated in the second dimension because of the larger frequency dispersion of the ^{13}C shifts. This proves to be particularly advantageous especially in the case of larger molecules, a feature illustrated by the identification of several natural products (problems 43-55). The resolution of overlapping *AB* systems as in the case of ring C*H$_2$* groups in steroids and in di- and triterpenes is especially helpful (problem 51). If there is sufficiently good resolution of the proton dimension in the spectrum, it may even be possible to recognise the fine structure of the ^1H multiplets from the shape of the correlation signals, a feature which is useful for solving problems 32, 48 and 51.

Figure 2.14 (page 37). C*H* COSY diagram of α-pinene [(CD$_3$)$_2$CO, 10 % v/v, 25 °C, 50 MHz for ^{13}C, 200 MHz for ^1H, 64 scans, 256 experiments]. (**a**) Complete contour plot; (**b**) stacked plot of the signals for the section outlined in (**a**) from $\delta_C = 20.9$ to 47.2 (^{13}C) and $\delta_H = 0.85$ to 2.34 (^1H); (**c**) contour plot of section (**b**), showing one-dimensional ^{13}C and ^1H NMR specta for this section aligned parallel to the abscissa and the ordinate

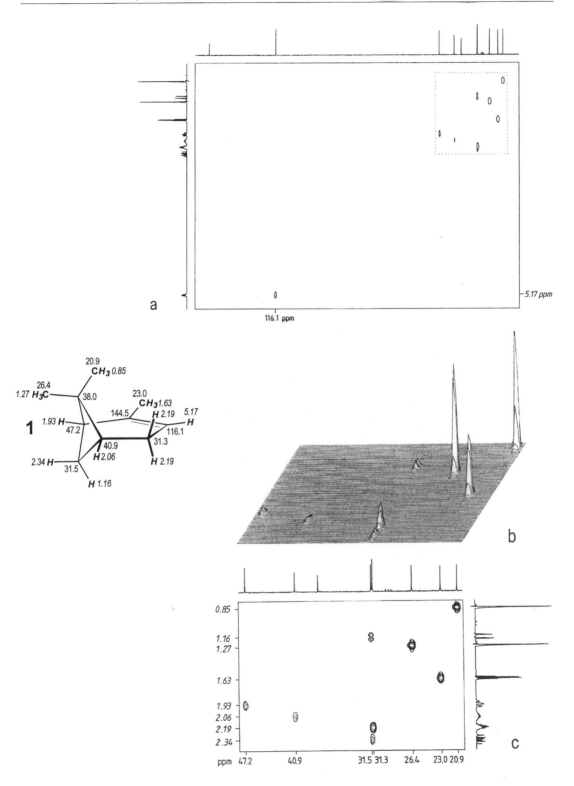

a

b

c

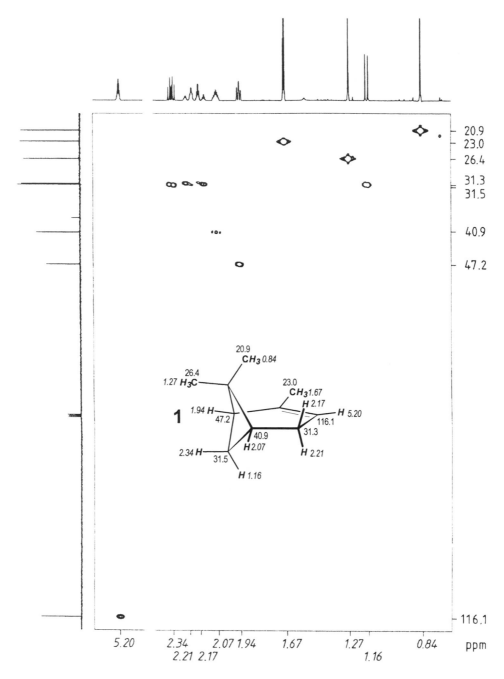

Figure 2.15. *HC* HSQC experiment (contour plot) of α-pinene [(CDCl₃, 5 % v/v, 25 °C, 125 MHz for ¹³C, 500 MHz for *¹H,* 4 scans, 256 experiments]. This experiment gives the same information as Fig. 2.14 within 8 minutes instead of two hours required for the *CH*-COSY in Fig. 2.14 due to higher sensitivity because of proton detection and stronger magnetic field. Deviations of proton shifts from those in Fig. 2.14 arise from the change of the solvent. The methylene protons collapsing in Fig. 2.14 at δ_H = 2.19 (200 MHz) display in this experiment an *AB* system with δ_A = 2.17 and δ_B = 2.21 (500 MHz)

2.2.9 Two-dimensional carbon-proton shift correlation via long-range C*H* coupling

Two-dimensional C*H* correlations such as C*H* COSY or *H*C HMQC and HSQC provide the $^1J_{CH}$ connectivities, and thereby apply only to those C atoms which are linked to *H* and not to non-protonated C atoms. Modifications of these techniques, also applicable to quaternary C atoms, are those which are adjusted to the smaller $^2J_{CH}$ and $^3J_{CH}$ couplings (2-25 Hz, Tables 2.8 and 2.9) [16,23]. Experiments that probe these couplings include the C*H* COLOC [24] (correlation *via* long range couplings) with carbon-13 detection (Fig. 2.16) and *H*C HMBC (heteronuclear multiple bond coherence) with the much more sensitive proton detection (Fig. 2.17) [17].

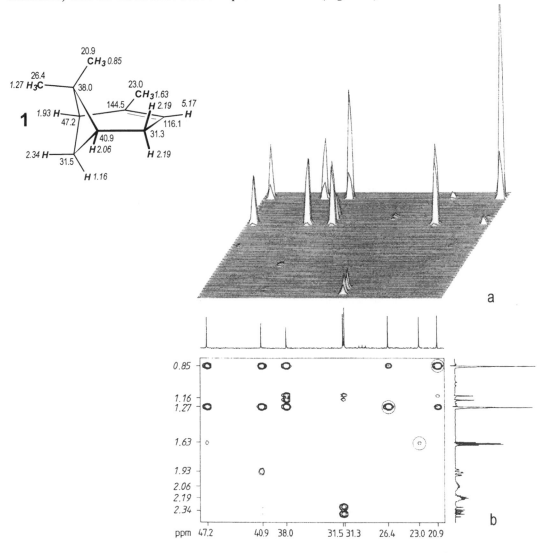

Figure 2.16. C*H* COLOC experiment of α-pinene [(CD$_3$)$_2$CO, 10% v/v, 25 °C, 50 MHz for ^{13}C, 200 MHz for ^{1}H, 256 scans and experiments]. (**a**) stacked plot of the section between δ_C = 20.9 and 47.2 (^{13}C) and δ_H = 0.85 and 2.34 (^{1}H); (**b**) contour plot of (**a**). One-dimensional ^{13}C and ^{1}H NMR spectra for this section are shown aligned with the abscissa and ordinate of the contour plot. $^1J_{CH}$ correlation signals which are already known from the C*H* COSY study (Fig. 2.14) and have not been suppressed, are indicated by circles

These two-dimensional C*H* shift correlations indicate C*H* relationships through two and more bonds (predominantly $^2J_{CH}$ and $^3J_{CH}$ connectivities) in addition to more or less suppressed $^1J_{CH}$ relationships which are in any case established from the C*H* COSY contour diagram. Format and analysis of the C*H* COLOC or HMBC plots correspond to those of a C*H* COSY or HSQC experiment, as is shown for α-pinene (**1**) in Figs. 2.14 - 2.17.

When trying to establish the C*H* relationships of a carbon atom (exemplified by the quaternary C at $\delta_C = 38.0$ in Figs. 2.16 and 2.17), the chemical shifts of protons at a distance of two or three bonds are found parallel to the proton axis (e.g. $\delta_H = 0.85\ (0.84)$, *1.16* and *1.27* in Figs. 2.16 and 2.17). It is also possible to take a proton signal as the starting point and from the cross signals parallel to the carbon axis to read off the shifts of the C atoms two or three bonds distant respectively. Thus, for example, one deduces that the methyl protons at $\delta_H = 0.84$ and *1.27* are two and three bonds apart from the C atoms at $\delta_C = 38.0$, 40.9 and 47.2 as illustrated by the partial structures **1a** and **1b** in Fig. 2.17. C*H* correlation signals due to methyl protons prove to be especially reliable, as do *trans* C*H* relationships over three bonds, e.g. between $\delta_H = 1.16$ and $\delta_C = 38.0$ in Figs. 2.16 and 2.17, in contrast to the missing *cis* relationship between $\delta_H = 2.34$ and $\delta_C = 38.0$ (partial structures **1d** and **1g** in Fig. 2.17) due to the smaller coupling constant to which the experiment was not adjusted.

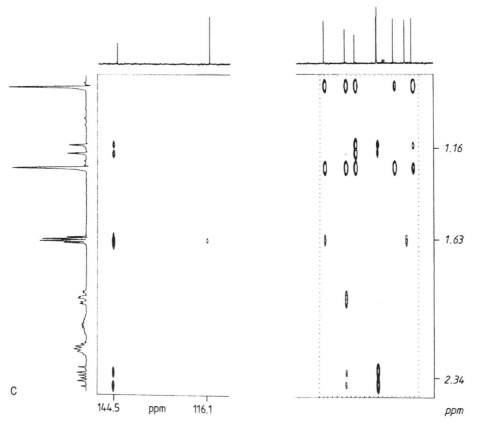

Figure 2.16 (c). *CH COLOC experiment of α-pinene [(CD$_3$)$_2$CO, 10 % v/v, 25 °C, 50 MHz for ^{13}C, 200 MHz for ^{1}H, 256 scans and experiments], complete contour plot (caption on previous page)*

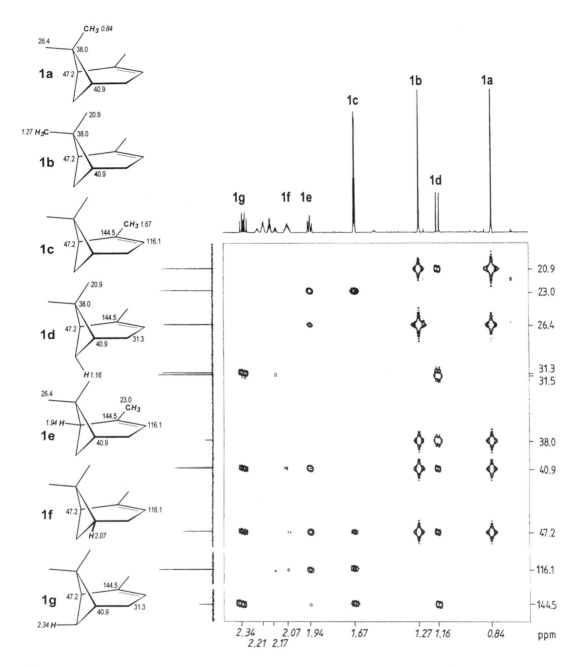

Figure 2.17. *HC* HMBC experiment of α-pinene [(CDCl₃, 5 % v/v, 25 °C, 125 MHz for ¹³C, 500 MHz for ¹H, 16 scans, 256 experiments, contour plot]. This experiment gives the same information as Fig. 2.16 within 24 min instead of 8 hrs required for the C*H*-COLOC in Fig. 2.16 due to higher sensitivity because of proton detection and stronger magnetic field. Deviations of proton shifts from those in Fig. 2.16 arise from the change of the solvent. The methylene protons collapsing in Fig. 2.16 at δ_H =2.19 (200 MHz) display in this experiment an *AB* system with   δ_A = 2.17 and δ_B = 2.21 (500 MHz)

2.3   Relative configuration and conformation

2.3.1   *HH* Coupling constants

Vicinal coupling constants $^3J_{HH}$ indicate very clearly the relative configuration of the coupling protons. Their contribution depends, according to the Karplus-Conroy equation [2,3],

$$^3J_{HH} = a \cos^2 \varphi - 0.28 \quad \text{(up to } \varphi = 90°: a \approx 10 \text{; above } \varphi = 90°: a \approx 15) \quad (2)$$

on the dihedral angle φ, enclosed by the *CH* bonds as shown in Fig. 2.18, which sketches the Karplus-Conroy curves for dihedral angles from 0 to 180°. Experimental values correspond to those given by the curve shown; deviations are up to *3 Hz*; electronegative substituents on the coupling path, for example, reduce the magnitude of $^3J_{HH}$.

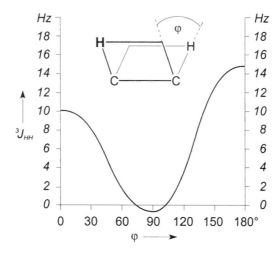

Figure 2.18. *Vicinal HH* coupling constants $^3J_{HH}$ as a function of the dihedral angle φ of the *CH* bonds concerned (Karplus-Conroy relationship)

For the stable conformers **13a-c** of a substituted ethane the *vicinal HH* coupling constants $J_s \approx 3\,Hz$ for *syn-protons* and $J_a \approx 15\,Hz$ for *anti-protons* can be derived from Fig. 2.18. If there is rotation around the C–C single bond, the coupling protons pass through the *syn* configuration twice and the *anti* configuration once.

$\varphi = 60°$	$\varphi = 180°$	$\varphi = -60°$
syn (gauche)	*anti (trans)*	*– syn (gauche)*
13a	**13b**	**13c**

Therefore, from

$$^3J_{(average)} = (^2J_s + J_a) / 3 = 21/3 = 7\,Hz \qquad (3)$$

an average coupling constant of about *7 Hz* is obtained. This coupling constant characterises alkyl groups with unimpeded rotation (cf. Figs 1.2 and 2.19).

Ethyl dibromodihydrocinnamate (**14**), for example, can form the three staggered conformers **14a-c** by rotation around the CC single bond α to the phenyl ring.

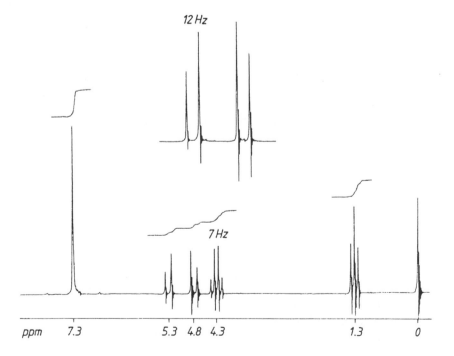

The ^1H NMR spectrum (Fig. 2.19) displays an *AB* system for the protons adjacent to this bond; the coupling constant $^3J_{AB} = 12\,Hz$. From this can be deduced first that the dihedral angle φ between the C*H* bonds is about 180°, second that conformer **14b** with minimised steric repulsion between the substituents predominates and third that there is restricted rotation around this CC bond.

Figure 2.19. ^1H NMR spectrum of ethyl dibromodihydrocinnamate (**14**) [CDCl$_3$, 25 °C, 90 MHz, CW recording]

The relative configuration of the protons which is deduced from the coupling constant $^3J_{AB}$ confirms the conformation of this part of the structure of this molecule. On the other hand, the $^3J_{HH}$ coupling constant of the ethyl group attached to oxygen (*7 Hz*, Fig. 2.19) reflects equal populations of all stable conformers around the CC bond of this ethyl group.

The $^3J_{HH}$ couplings shown in Table 2.10 verify the Karplus-Conroy equation 2 (Fig. 2.18) for rigid systems. Hence in cyclopropane the relationship $^3J_{HH(cis)} > {}^3J_{HH(trans)}$ holds, because *cis*-cyclopropane protons enclose a dihedral angle of about 0°, in contrast to an angle of *ca* 145° between *trans* protons, as shown by Dreiding models. *Vicinal* protons in cyclobutane, cyclopentane, norbornane and norbornene behave in an analogous way with larger *cis, endo-endo* and *exo-exo* couplings, respectively (Table 2.10).

Substituent effects (electronegativity, configuration) influence these coupling constants in four-, five- and seven-membered ring systems, sometimes reversing the *cis-trans* relationship [2,3], so that other NMR methods of structure elucidation, e.g. NOE difference spectra (see Section 2.3.5), are needed to provide conclusive results. However, the coupling constants of *vicinal* protons in cyclohexane and its heterocyclic analogues (pyranoses, piperidines) and also in alkenes (Table 2.10) are particularly informative.

Neighbouring *diaxial* protons of cyclohexane can be clearly identified by their large coupling constants ($^3J_{aa} \approx 11\text{-}13\ Hz$, Table 2.10) which contrast with those of protons in *diequatorial* or *axial-equatorial* configurations ($^3J_{ee} \approx {}^3J_{ae} \approx 2\text{-}4\ Hz$). Similar relationships hold for pyranosides as oxygen hetero analogues of cyclohexane, wherein the electronegative *O* atoms reduce the magnitude of the coupling constants ($^3J_{aa} \approx 9\ Hz$, $^3J_{ae} \approx 4\ Hz$, Table 2.10). These relationships are used for elucidation of the configuration of substituted cyclohexanes (problem 37), cyclohexenes (problems 12 and 35), terpenes (problems 47, 48, 50 and 51), flavans (problem 10) and glycosides (problem 45, Table 2.10). In these cases also, the relative configuration of the protons which is deduced from the $^3J_{HH}$ coupling constant reveals the conformation of the six-membered rings. Thus the coupling constant *9 Hz* of the protons in positions 1 and 2 of the methyl-β-D-glucopyranoside **15** determines not only the *diaxial* configuration of the coupling protons but also the $^4C_1$ conformation of the pyranose ring. If the sterically more crowded $^1C_4$ conformation (exclusively with axial substituents) were present then a *diequatorial* coupling (*4 Hz*) of protons *1-H* and *2-H* would be observed. If the conformers were inverting (50 : 50 population of the $^4C_1$ and $^1C_4$ conformers), then the coupling constant would be the average (*6.5 Hz*).

The couplings of *vicinal* protons in 1,2-disubstituted alkenes lie in the range *6-12 Hz* for *cis* protons (dihedral angle 0°) and *12-17 Hz* for *trans* protons (dihedral angle 180°), thus also following the Karplus-Conroy equation. Typical examples are the alkene proton *AB* systems of coumarin (**16a**, *cis*) and *trans*-cinnamic acid (**16b**), and of the *cis-trans* isomers **17a** and **b** of ethyl isopentenyl ether, in addition to those in problems 3, 4, 8, 11, 13 and 38.

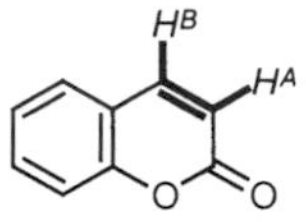

16a	16b	17a	17b
$^3J_{AB}$ = 9.5 Hz (cis)	$^3J_{AB}$ = 15.5 Hz (trans)	$^3J_{AB}$ = 6 Hz (cis)	$^3J_{AB}$ = 13 Hz (trans)

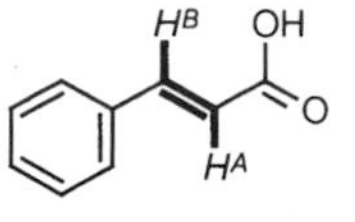

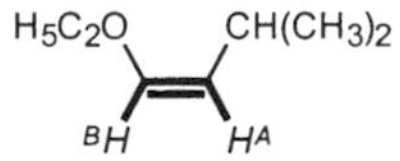

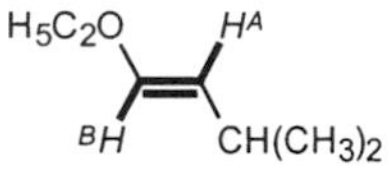

Table 2.10. $^3J_{HH}$ coupling constants (*Hz*) and relative configuration [2,3]. The coupling path is shown in bold

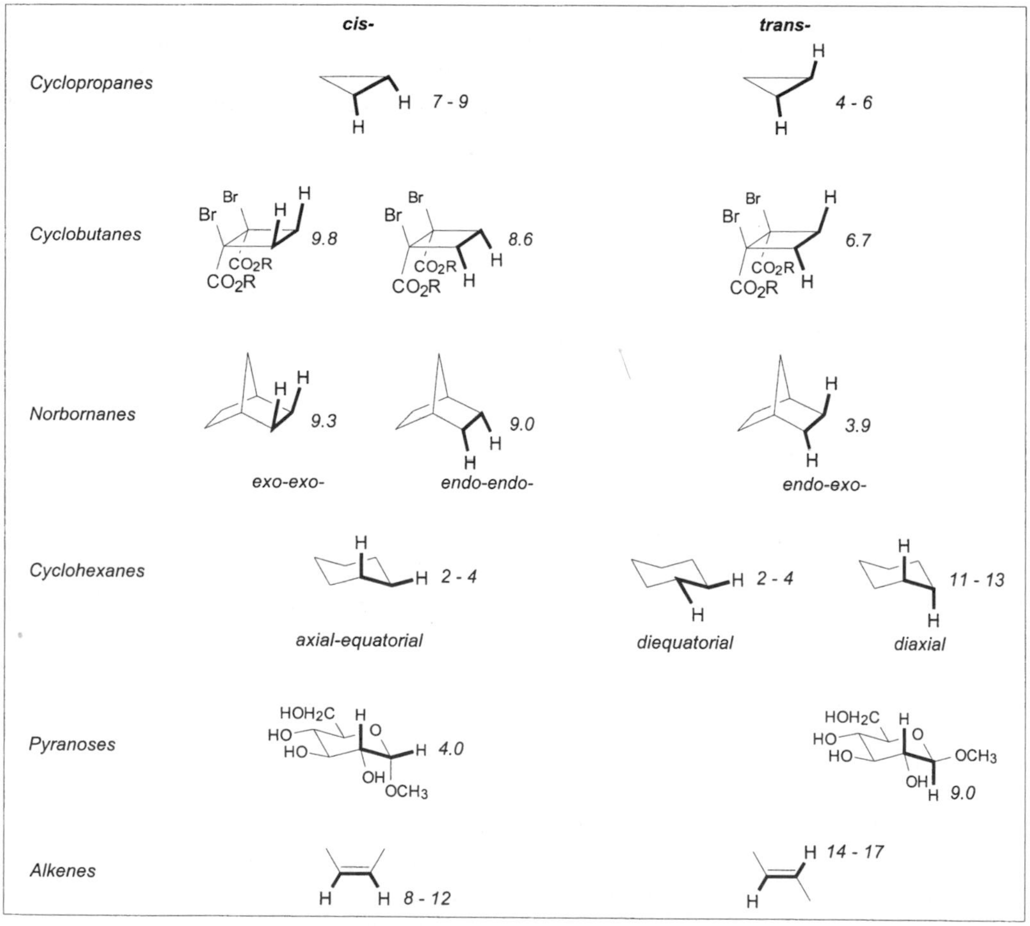

2.3.2   *CH* Coupling constants

Geminal CH coupling constants $^2J_{CH}$ reflect the configuration of electronegative substituents in molecules with a defined geometry such as pyranose and alkenes [16]. If an electronegative substituent is attached *cis* with respect to the coupling proton, then the coupling constant $^2J_{CH}$ has a higher negative value; if it is located *trans* to the coupling proton, then $^2J_{CH}$ is positive and has a lower value; this is illustrated by β- and α-D-glucopyranose (**18a** and **b**) and by bromoethene (**19**).

18a   **18b**   **19**

Vicinal CH coupling constants $^3J_{CH}$ resemble *vicinal HH* coupling constants in the way that they depend on the cosine^2 of the dihedral angle φ between the CC bond to the coupled C atom and the *CH* bond to the coupled proton [16] (cf. Fig. 2.16), as illustrated by the Newman projections of the conformers **20a-c** of a propane fragment.

It follows from this that where there is equal population of all conformers in an alkyl group (free rotation about the CC single bond) then an averaged coupling constant $^3J_{CH} = (2J_{syn} + J_{anti})/3$ of between 4 and 5 Hz can be predicted, and that *vicinal CH* coupling constants $^3J_{CH}$ have values about two thirds of those of *vicinal* protons, $^3J_{HH}$ [16].

Like $^3J_{HH}$ couplings, $^3J_{CH}$ couplings give conclusive information concerning the relative configuration of C and *H* as coupled nuclei in cyclohexane and pyranose rings and in alkenes (Table 2.11). Substituted cyclohexanes have $^3J_{CH} \approx 2$-4 Hz for *cis* and 8-9 Hz for *trans* configurations of the coupling partners; electronegative *OH* groups on the coupling path reduce the magnitude of $^3J_{CH}$ in pyranoses (Table 2.11). When deducing the configurations of multi-substituted alkenes, e.g. in solving problem 19, the $^3J_{CH}$ couplings of the alkenes in Table 2.11 are useful.

$^3J_{CH(trans)} > ^3J_{CH(cis)}$ holds throughout. Electronegative substituents on the coupling carbon atom increase the *J*-value, whilst reducing it on the coupling path. Moreover, $^3J_{CH}$ reflects changes in the bonding state (carbon hybridisation) and also steric hindrance (impeding coplanarity), as further examples in Table 2.11 show.

Table 2.11. $^3J_{CH}$ coupling constants (Hz) and relative configuration [16]. The coupling path is shown in bold

2.3.3 *NH* Coupling constants

The relationship between $^3J_{NH}$ and the dihedral angle of the coupling nuclei, of the type that applies to *vicinal* couplings of ^1H and ^{13}C, very rarely permits specific configurational assignments because the values ($^3J_{NH} < 5$ Hz) are too small [7]. In contrast, *geminal* couplings $^2J_{NH}$ distinguish the relative configurations of aldimines very clearly. Thus, *anti*-furan-2-aldoxime (**21a**) shows a considerably larger $^2J_{NH}$ coupling than does the *syn* isomer **21b**; evidently in imines the non-bonding electron pair *cis* to the *CH* bond of the coupled proton has the effect of producing a high negative contribution to the *geminal NH* coupling.

2.3.4 ^{13}C Chemical shifts

A carbon atom in an alkyl group is shielded by a substituent in the γ-position, that is, it experiences a smaller ^{13}C chemical shift or a negative substituent effect [4-6]. This originates from a sterically induced polarisation of the CH bond: the van der Waals radii of the substituent and of the hydrogen atom on the γ-C overlap; as a result, the σ-bonding electrons are moved from H towards the γ-C atom; the higher electron density on this C atom will cause shielding. As the Newman projections **22a-c** show, a distinction can be made between the stronger γ-*syn* and the weaker γ-*anti* effect. If there is free rotation, then the effects are averaged according to the usual expression, $(2\gamma_{syn}+\gamma_{anti})/3$, and one observes a negative γ-substituent effect of –2.5 to –3.5 ppm [4-6], which is typical for alkyl groups.

22a   **22b**   **22c**

In rigid molecules, strong γ-effects on the ^{13}C shift (up to 10 ppm) allow the different configurational isomers to be distinguished unequivocally, as *cis*- and *trans*-3- and 4-methylcyclohexanol (Table 2.12) illustrate perfectly: if the OH group is positioned *axial*, then its van der Waals repulsion of a *coaxial H* atom shields the attached C atom in the γ-position. *1,3-Diaxial* relationships between substituents and H atoms in cyclohexane, norbornane and pyranosides shield the affected C atoms, generating smaller ^{13}C shifts than for isomers with *equatorial* substituents (Table 2.12).

The ^{13}C chemical shift thus reveals the relative configuration of substituents in molecules with a definite conformation, e.g. the *axial* position of the OH group in *trans*-3-methylcyclohexanol, *cis*-4-methylcyclohexanol, β-D-arabinopyranose and α-D-xylopyranose (Table 2.12, p. 50). It turns out, in addition, that these compounds also take on the conformations shown in Table 2.12 (arabinopyranose, $^1C_4$; the others, $^4C_1$); if they occurred as the other conformers, then the OH groups on C-1 in these molecules would be *equatorial* with the result that larger shifts for C-l, C-3 and C-5 would be recorded. A ring inversion (50 : 50 population of both conformers) would result in an average ^{13}C shift.

Compared with ^1H chemical shifts, ^{13}C shifts are more sensitive to steric effects, as a comparison of the ^1H and the ^{13}C NMR spectra (*cis*- and *trans*-4-*tert*-butylcyclohexanol **23**) in Fig. 2.20 shows. The polarisation through space of the γ-CH bond by the *axial* OH group in the *cis* isomer **23b** shields C-l by –5.6 and C-3 by –4.8 ppm (γ-effect). In contrast the ^1H shifts reflect the considerably smaller anisotropic effect (see Section 2.5.1) of cyclohexane bonds: *equatorial* substituents (in this case H and OH) display larger shifts than *axial* substituents; the *equatorial* 1-H in the *cis*-isomer **23b** (δ_H=3.92) has a larger shift than the *axial* 1-H in the *trans*-compound **23a** (δ_H=3.40); the difference is significantly smaller (*–0.52 ppm*) than the γ-effect on the ^{13}C shifts (*ca* –5 ppm). Both spectra additionally demonstrate the value of NMR spectroscopy for quantitative analysis of mixtures by measuring integral levels or signal intensities, respectively. Finally, D$_2$O exchange eliminates the OH protons and their couplings from the ^1H NMR spectra (Fig. 2.20d).

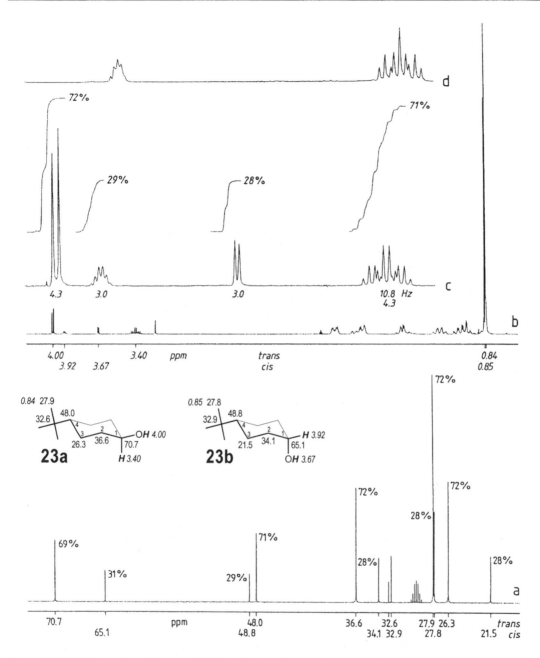

Figure 2.20. NMR spectra of *trans*-and *cis*-4-*tert*-butylcyclohexanol (**23a** and **23b**) [(CD₃)₂CO, 25 °C, 400 MHz for ¹H, 100 MHz for ¹³C]. (**a**) ¹H decoupled ¹³C NMR spectrum (NOE suppressed, comparable signal intensities); (**b**) ¹H NMR spectrum; (**c**) section of (**b**) (δ_H = 3-4) with integrals; (**d**) partial spectrum (**c**) following D₂O exchange. The integrals (**c**) and the ¹³C signal intensities (**a**) give the *trans* : *cis* isomer ratio 71 : 29. Proton 1-*H* (δ_H = 3.40) in the *trans* isomer **23a** forms a triplet (10.8 Hz, two *anti* protons in 2,2'-positions) of quartets (4.3 Hz, two *syn* protons in 2,2'-positions and the O*H* proton as additional coupling partner); following D₂O exchange a triplet (10.8 Hz) of triplets (4.3 Hz) appears, because the coupling to O*H* is missing. In the *cis* isomer **23b** proton 1-*H* forms a sextet (3.0 Hz, four *synclinal* protons in 2,2'-positions and O*H*) which appears as a quintet following D₂O exchange because the coupling to O*H* is then lost

Table 2.12. ^{13}C chemical shifts (δ_C) and relative configurations of cycloalkanes, pyranoses and alkenes (application of γ-effects) [4-6]. The shifts which are printed in boldface reflect γ-effects on C atoms in the corresponding isomer pairs

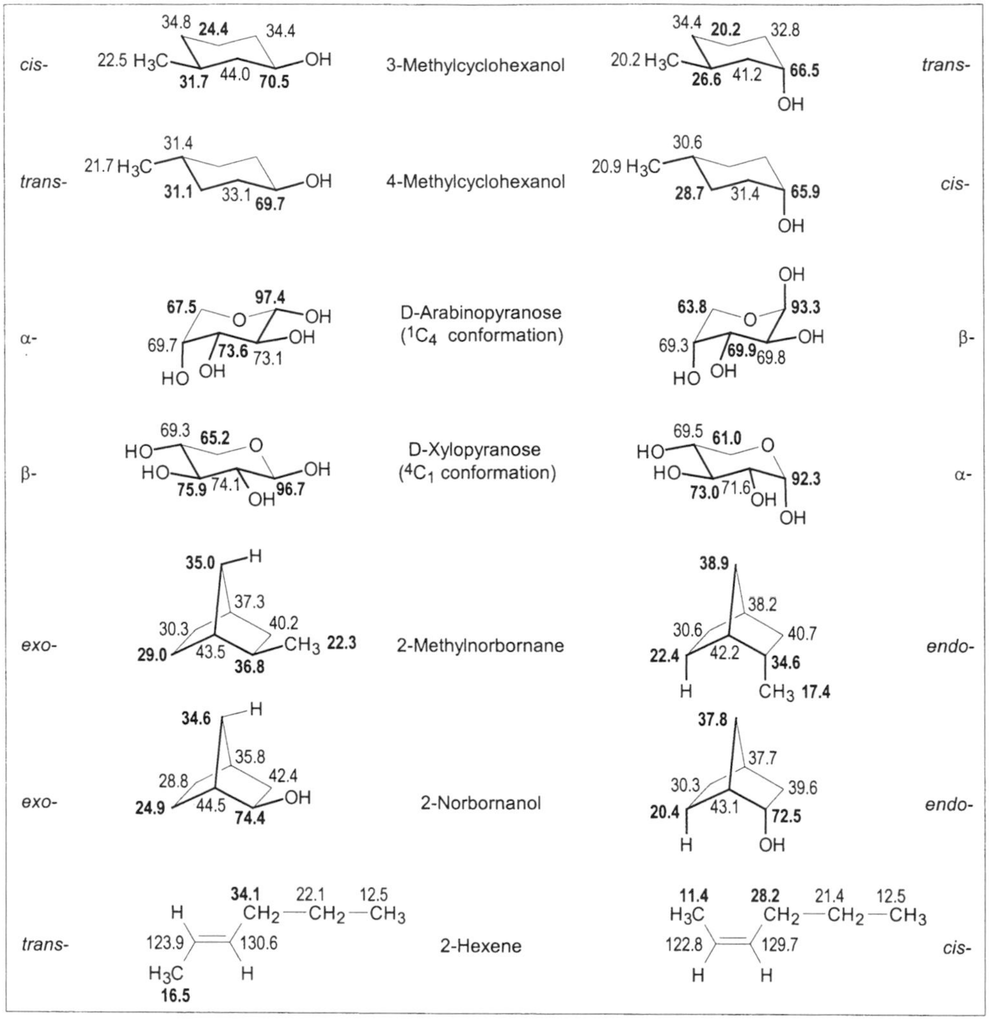

The γ-effect on the ^{13}C shift also causes the difference between (*E*)- and (*Z*)-configurations of the alkyl groups in alkenes. Here the C-α atom shift responds most clearly to the double bond configurational change: these atoms in *cis*-alkyl groups occupy γ-positions with respect to each other; they are eclipsed, enclosing a dihedral angle of 0°, which leads to an especially strong van der Waals interaction and a correspondingly strong shielding of the ^{13}C nucleus. For this reason, the relationship [$\delta_{trans} > \delta_{cis}$] holds for the α-C atoms of alkenes, as shown in Table 2.12 for (*E*)- and (*Z*)-2-hexene. The ^{13}C shifts of the doubly bonded carbon atoms behave similarly, although the effect is considerably smaller.

α,β-Unsaturated carbonyl compounds show smaller ^{13}C shifts than comparable saturated compounds [4-6], provided that their carbonyl and CC double bonds are coplanar. If steric hindrance prevents coplanarity, conjugation is reduced and so larger ^{13}C shifts are observed. In α,β-unsaturated carbonyl compounds such as benzophenones and benzoic acid derivatives the twist angle θ between the carbonyl double bond and the remaining π-system can be read off and hence the conformation derived from the ^{13}C shift [25], as several benzoic acid esters (**24**) illustrate.

R	$\delta_{C=O}$	$\theta\ °$
H	166.9	0
CH_3	170.4	49
$CH(CH_3)_2$	171.3	57
$C(CH_3)_3$	173.1	90

2.3.5   NOE Difference spectra

Changes in signal intensities caused by spin decoupling (double resonance) are referred to as the nuclear Overhauser effect (NOE) [3,26]. In proton decoupling of ^{13}C NMR spectra, the NOE increases the intensity of the signals generated by the C atoms which are bonded to hydrogen by up to 300 %; almost all techniques for measuring ^{13}C NMR spectra exploit this gain in sensitivity [2-6]. If in recording ^{1}H NMR spectra certain proton resonances are decoupled (homonuclear spin decoupling), then the changes in intensity due to the NOE are considerably smaller (much less than 50 %).

For the assignment of configuration it is useful that, during perturbing the equilibrium population of a particular proton by irradiating it for long enough, other protons in the vicinity may be affected although not necessarily coupled with this proton. As a result of molecular motion and the dipolar relaxation processes associated with it, the populations of energy levels of the protons change [3,26]; their signal intensities change accordingly (NOE). For example, if the signal intensity of one proton increases during irradiation of another, then these protons must be positioned close to one another in the molecule, irrespective of the number of bonds which separate them.

NOE difference spectroscopy has proved to be a useful method for studying the spatial proximity of protons in a molecule [27]. In this experiment the ^{1}H NMR spectrum is recorded during the irradiation of a particular proton (measurement 1); an additional measurement with an irradiation frequency which lies far away (the 'off-resonance' experiment) but is otherwise subject to the same conditions, is then the basis for a comparison (reference measurement 2). The difference between the two measurements provides the NOE difference spectrum, in which only those signals are shown whose intensities are increased (positive signal) or decreased (negative signal) by NOE. Figure 2.21 illustrates NOE difference spectroscopy with α-pinene (**1**): irradiation of the methyl protons at $\delta_H = 1.27$ (experiment **c**) gives a significant NOE on the proton at $\delta_H = 2.34$; if for comparison, the methyl protons with $\delta_H = 0.85$ are perturbed (experiment **b**), then no NOE is observed at $\delta_H = 2.34$. From this the proximity of the methylene-H atom at $\delta_H = 2.34$ and the methyl group at 1.27 in α-pinene is detected. In addition, both experiments confirm the assignment of the methyl protons to the signals at $\delta_H = 0.85$ and 1.27. A negative NOE, as on the protons at $\delta_H = 1.16$ in experiment **c**, is the result of coupling, e.g. in the case of the *geminal* relationship with the affected proton at $\delta_H = 2.34$. Further applications of NOE difference spectroscopy are provided in problems 29, 33, 34, 36, 44, 48, and 50-54.

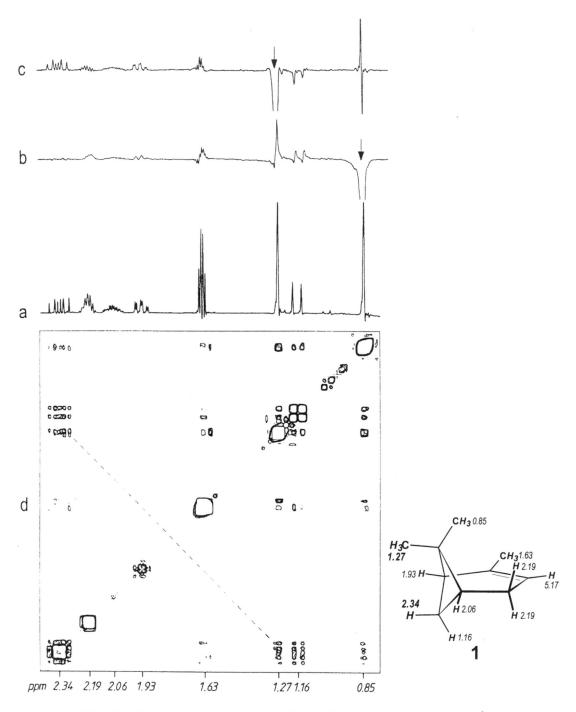

Figure 2.21. *HH* NOE difference spectra (**b**, **c**) and *HH* NOESY diagram (**d**) of α-pinene (**1**) with ^1H NMR spectrum (**a**) for comparison [(CD₃)₂CO, 10% v/v, 25 °C, 200 MHz, section from $\delta_H = 0.85$ to 2.34]. Vertical arrows in (**b**) and (**c**) indicate the irradiation frequencies; in the *HH* NOESY plot (**d**), cross-signals linked by a dotted line show the NOE detected in (**c**)

2.3.6 *HH* NOESY and ROESY

The *HH* COSY sequence for ascertaining the *HH* connectivities also changes the populations of the energy levels, leading to NOEs. Thus, the *HH* COSY experiment has been modified to pulse sequences such as *HH* NOESY [17,28] which allows a two-dimensional detection of NOEs. The result of such measurements is shown in the *HH* NOESY plot with square symmetry (Fig. 2.21**d**) which is evaluated in the same way as *HH* COSY. Fig. 2.21**d** shows by the cross signals at δ_H = 2.34 and 1.27 that the appropriate protons in α-pinene (**1**) are close to one another; the experiment also illustrates that the *HH* COSY cross signals (due to through-bond coupling) are not completely suppressed. Therefore, before evaluating a two-dimensional NOE experiment, it is essential to know the *HH* connectivities from the *HH* COSY plot.

The NOESY sequence sensitively reacts towards the change of the sign of NOEs depending on molecular motion. The ROESY experiment [2,17] involving an isotropic mixture of magnetisations by spin-locking detects NOEs without these limitations. Thus, the proximity of the methyl protons in α-pinene ((δ_H = 0.84 in CDCl$_3$) to the *exo*-methylene proton (δ_H = 2.17) is additionally indicated and assigned in the ROESY experiment (Fig. 2.22). Problems 46 and 55 exemplify further applications.

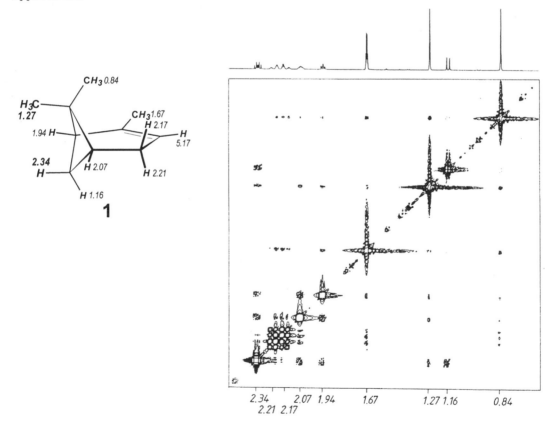

Figure 2.22. *HH* ROESY diagram of α-pinene (**1**) with ^1H NMR spectrum [CDCl$_3$, 10% v/v, 25 °C, 500 MHz, section from δ_H = 0.84 to 2.34]. Deviations of chemical shifts from those in other experiments (Figs. 2.14, 2.16) arise from solvent effects; the methylene protons collapsing in Fig. 2.21 at δ_H = 2.19 (200 MHz) display in this experiment an *AB* system with δ_A = 2.17 and δ_B = 2.21 (500 MHz)

A comparison of the methods of proton-proton NOE detection has shown that two-dimensional NOE detection such as NOESY and ROESY are better suited to the investigation of the stereochemistry of biopolymers whereas for small- to medium-sized molecules (up to 30 C atoms) NOE difference spectroscopy is less time consuming, more selective and thus more conclusive.

2.4 Absolute configuration

2.4.1 Diastereotopism

Where both H atoms of a methylene group cannot be brought into a chemically identical position by rotation or by any other movement of symmetry, they are said to be *diastereotopic* [2,3]. The precise meaning of diastereotopism is best illustrated by means of an example, that of methylene protons H^A and H^B of glycerol (**25**). Where there is free rotation about the CC bonds, the terminal CH_2OH groups rotate through three stable conformations. They are best shown as Newman projections (**25a-c**) and the chemical environments of the CH_2O protons, H^A and H^B are examined with particular reference to *geminal* and *synclinal* neighbours.

25a	**25b**	**25c**

H^A : OH , H^B ; OH , CH_2OH : δ_1 OH , H^B ; CH_2OH , H^C : δ_2 OH , H^B ; H^C , OH : δ_3

H^B : H^A , OH ; CH_2OH , H^C : δ_4 H^A , OH ; H^C , OH : δ_5 H^A , OH ; OH , CH_2OH : δ_6

It can be seen that the six possible near-neighbour relationships are all different. If rotation were frozen, then three different shifts would be measured for H^A and H^B in each of the conformers **a, b** and **c** (δ_1, δ_2 and δ_3 for H^A, δ_4, δ_5 and δ_6 for H^B). If there is free rotation at room temperature and if χ_a, χ_b and χ_c are the populations of conformers **a, b** and **c**, then according to the equations 4,

$$\delta_A = \chi_a\delta_1 + \chi_b\delta_2 + \chi_c\delta_3 \quad \text{and} \quad \delta_B = \chi_a\delta_4 + \chi_b\delta_5 + \chi_c\delta_6 \qquad (4)$$

different average shifts δ_A and δ_B are recorded which remain differentiated when all three conformations occur with equal population ($\chi_a = \chi_b = \chi_c = 1/3$). Chemical equivalence of such protons would be purely coincidental.

Figure 2.23 shows the diastereotopism of the methylene protons (CH^AH^BOD) of glycerol (**25**) in D_2O solution (*OH* exchanged to *OD*); it has a value of $\delta_B - \delta_A = 0.09$. The spectrum displays an $(AB)_2C$ system for the symmetric constitution, $(CH^AH^BOD)_2CH^COD$, of the molecule with *geminal* coupling $^2J_{AB} = 11.6\ Hz$ and the *vicinal* coupling constants $^3J_{AC} = 6.4$ and $^3J_{BC} = 4.5\ Hz$. The unequal ^3J couplings provide evidence against the unhindered rotation about the CC bonds of glycerol and indicate instead that conformer **a** or **c** predominates with a smaller interaction of the substituents compared to **b**.

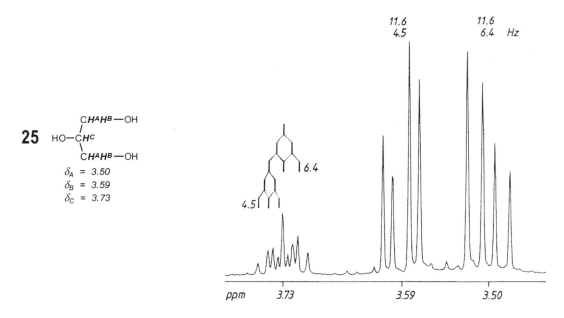

Figure 2.23. ^1H NMR spectrum of glycerol [**25**, D_2O, 10 %, 25 °C, 400 MHz]

Diastereotopism indicates *prochirality*, as exemplified by glycerol (**25**, Fig. 2.23). Other examples of this include diethylacetals, in which the OCH_2 protons are diastereotopic on account of the prochiral acetal-C atoms, thus forming *AB* systems of quartets because of coupling with the methyl protons.

The Newman projections **25a-c** draw attention to the fact that the central C atom, as seen from the terminal CH_2OD groups, appears asymmetric. It follows from this that diastereotopism is also a way of probing neighbouring asymmetric C atoms. Thus the methyl groups of the isopropyl residues in *D*- or *L*-valine (**26**) are diastereotopic and so show different ^1H and ^{13}C shifts, although these cannot be individually assigned to the two groups. In chiral alcohols of the type **27** the diastereotopism of the isopropyl-C nucleus increases with the size of the alkyl residues (methyl < isopropyl < *tert*-butyl) [29].

If a molecule contains several asymmetric C atoms, then the diastereomers show diastereotopic shifts. Clionasterol (**28a**) and sitosterol (**28b**) for example, are two steroids that differ only in the absolute configuration at one carbon atom, C-24 [30]. Differing shifts of ^{13}C nuclei close to this asymmetric C atom in **28a** and **b** identify the two diastereomers including the absolute configuration of C-24 in both. The absolute configurations of carboxylic acids in pyrrolizidine ester alkaloids are also reflected in diastereotopic ^1H and ^{13}C shifts [31], which is used in solving problem 54.

28a Clionasterol (24*S*)

28b Sitosterol (24*R*)

2.4.2   Chiral shift reagents (ee determination)

The presence of asymmetric C atoms in a molecule may, of course, be indicated by diastereotopic shifts and absolute configurations may, as already shown, be determined empirically by comparison of diastereotopic shifts [30,31]. However, enantiomers are not differentiated in the NMR spectrum. The spectrum gives no indication as to whether a chiral compound exists in a racemic form or as a pure enantiomer.

Nevertheless, it is possible to convert a racemic sample with chiral reagents into diastereomers or simply to dissolve it in an enantiomerically pure solvent *R* or *S*; following this process, solvation diastereomers arise from the racemate (*RP* + *SP*) of the sample P, e.g. *R*:*RP* and *R*:*SP* , in which the enantiomers are recognisable because of their different shifts. Compounds with groups which influence the chemical shift because of their anisotropy effect (see Sections 2.5.1 and 2.5.2) are suitable for use as chiral solvents, e.g. 1-phenylethylamine and 2,2,2-trifluoro-1-phenylethanol [32].

A reliable method of checking the enantiomeric purity by means of NMR uses europium(III) or praseodymium(III) chelates of type **29** as chiral shift reagents [33]. With a racemic sample, these form diastereomeric europium(III) or praseodymium(III) chelates, in which the shifts of the two enantiomers are different. Different signals for *R* and *S* will be observed only for those nuclei in immediate proximity to a group capable of coordination (O*H*, N*H₂*, C=O). The separation of the signals increases with increasing concentration of the shift reagent; unfortunately, line broadening of signals due to the paramagnetic ion increases likewise with an increase in concentration, which limits the amount of shift reagent which may be used. Figure 2.24 shows the determination of an enantiomeric excess (*ee*) following the equation 5

$$ee = \frac{R-S}{R+S} \times 100\,(\%) \qquad (5)$$

for 1-phenylethanol (**30**) by ^1H and ^{13}C NMR, using tris[3-(heptafluoropropylhydroxymethylene)-D-camphorato]praseodymium(III) (**29b**) as a chiral shift reagent.

	R	M
29a	CF_3	Eu^{3+}
29b	$CF_2-CF_2-CF_3$	Pr^{3+}

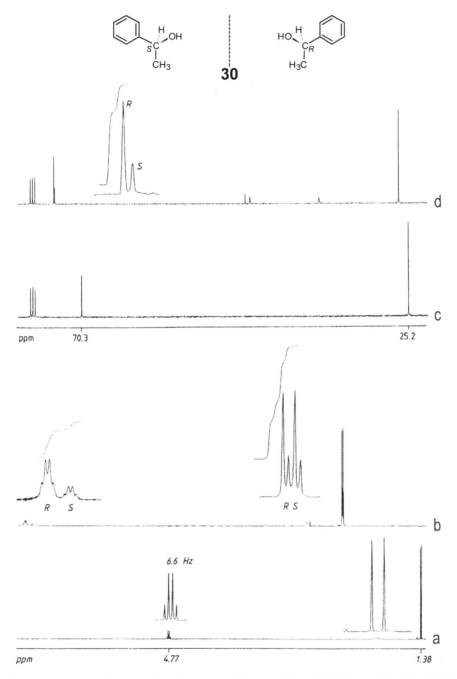

Figure 2.24. Determination of the enantiomeric excess of 1-phenylethanol [**30**, 0.1 mmol in 0.3 ml CDCl$_3$, 25 °C] by addition of the chiral praseodymium chelate **29b** (0.1 mmol). (**a, b**) ^1H NMR spectra (400 MHz), (**a**) without and (**b**) with the shift reagent **29b**. (**c, d**) ^{13}C NMR spectra (100 MHz), (**c**) without and (**d**) with the shift reagent **29b**. In the ^{13}C NMR spectrum (**d**) only the C-α atoms of enantiomers **30R** and **30S** are resolved. The ^1H and ^{13}C signals of the phenyl residues are not shifted; these are not shown for reasons of space. The evaluation of the integrals gives 73 % *R* and 27 % *S*, i.e. an enantiomeric excess (*ee*) of 46 %

2.5   Intra- and intermolecular interactions

2.5.1   Anisotropic effects

The chemical shift of a nucleus depends in part on its spatial position in relation to a bond or a bonding system. The knowledge of such anisotropic effects [2,3] is useful in structure elucidation. An example of the anisotropic effect would be the fact that *axial* nuclei in cyclohexane almost always show smaller ^1H shifts than *equatorial* nuclei on the same C atom (illustrated in the solutions to problems 37, 47, 48, 50 and 51). The γ-effect also contributes to the corresponding behaviour of ^{13}C nuclei (see Section 2.3.4).

Multiple bonds are revealed clearly by anisotropic effects. Textbook examples include alkynes, shielded along the C≡C triple bond, and alkenes and carbonyl compounds, where the nuclei are deshielded in the plane of the C=C and C=O double bonds, respectively [2,3]. One criterion for distinguishing methyl groups attached to the double bond of pulegone (**31**), for example, is the carbonyl anisotropic effect.

2.5.2   Ring current of aromatic compounds

Benzene shows a considerably larger ^1H shift ($\delta_H = 7.28$) than alkenes (cyclohexene, $\delta_H = 5.59$) or cyclically conjugated polyenes such as cyclooctatetraene ($\delta_H = 5.69$). This is generally explained by the deshielding of the benzene protons by a ring current of π-electrons [2,3] which is induced when an aromatic compound is subjected to a magnetic field. The ring current itself produces its own magnetic field, opposing the external field within and above the ring, but aligned with it outside [2,3]. As a result, nuclei inside or above an aromatic ring display a smaller shift whereas nuclei outside the ring on a level with it show a larger shift. The ring current has a stronger effect on the protons attached to or in the ring than on the ring C atoms themselves, so that particularly ^1H shifts prove a useful means of detecting ring currents and as aromaticity criteria for investigating annulenes.

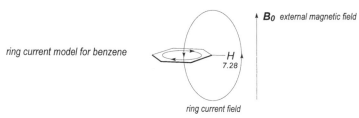

ring current model for benzene

1,4-Decamethylenebenzene (**32**) illustrates the ring current of benzene by a shielding of the methylene protons ($\delta_H = 0.8$) which lie above the aromatic ring plane in the molecule. A clear representation of the ring current effect is given by [18]annulene (**33**) at low temperature and the vinylogous porphyrin **35** with a diaza[26]annulene perimeter [34]: the inner protons are strongly shielded ($\delta_H = -2.88$ and -11.64, respectively); the outer protons are strongly deshielded ($\delta_H = 9.25$ and 13.67, respectively). The typical shift of the inner NH protons ($\delta_H = -2$ to -3) indicates that porphyrin **34** and its expanded analogues such as **35** occur as diaza[18]annulene tautomers. Problem 39 draws attention to this point.

2.5.3 Intra- and intermolecular hydrogen bonding

Hydrogen bonding can be recognised in ^1H NMR spectra by the large shifts associated with it [2,3]; these large shifts are caused by the electronegativity of the heteroatoms bridged by the hydrogen atom. The OH protons of enol tautomers of 1,3-diketones are extreme examples. They form an *intramolecular H* bond and appear between $\delta_H = 12.5$ (hexafluoroacetylacetone enol, Fig. 2.25) and 15.5 (acetylacetone enol).

Intermolecular H bonding can be recognised in the ^1H NMR spectrum by the fact that the shifts due to the protons concerned depend very strongly on the concentration, as the simple case of methanol (**36**) demonstrates (Fig.2.25**a**); solvation with tetrachloromethane as a solvent breaks down the hydrogen bridging increasingly with dilution of the solution; the OH shift decreases in proportion to this. In contrast, the shift of the ^1H signal of an intramolecular bridging proton remains almost unaffected if the solution is diluted as illustrated in the example of hexafluoroacetylacetone (**37**), which is 100 % enolised (Fig. 2.25**b**).

Intermolecular H bonding involves an exchange of hydrogen between two heteroatoms in two different molecules. The H atom does not remain in the same molecule but is exchanged. If its exchange frequency is greater than is given by the Heisenberg uncertainty principle,

$$\nu_{exchange} \geqq \pi J_{AX}/\sqrt{2} \sim 2.22\, J_{AX} \qquad (6)$$

then its coupling J_{AX} to a *vicinal* proton H^A is not resolved. Hence CH_n protons do not generally show splitting by *vicinal* SH, OH or NH protons at room temperature. The same holds for $^3J_{CH}$ couplings with such protons. If the hydrogen bonding is hindered by solvatation or is intramolecular, then coupling is resolved, as the example of salicylaldehyde (**10**) has already shown (see Section 2.2.4; for applications, see problems 17 and 25).

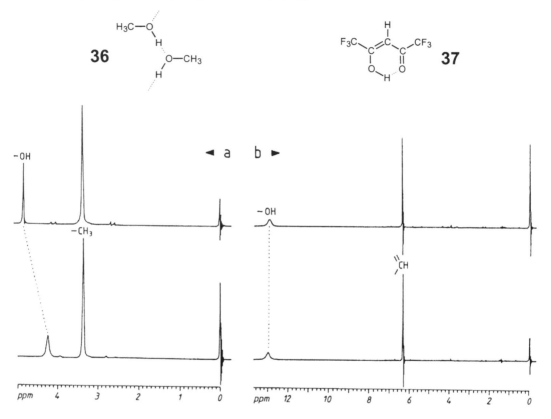

Figure 2.25. ^1H NMR spectra of methanol (**36**, **a**) and hexafluoroacetylacetone (**37**, **b**), both in the pure state (above) and diluted in tetrachloromethane solution (5%, below) [25 °C, 90 MHz, *CW* recording]

2.5.4   Protonation effects

If a sample contains groups that can take up or lose a proton, H^+ (NH_2, $COOH$), then one must expect the pH and the concentration to affect the chemical shift when the experiment is carried out in an acidic or alkaline medium to facilitate dissolution. The pH may affect the chemical shift of more distant, nonpolar groups, as shown by the amino acid alanine (**38**) in neutral (betaine form **38a**) or alkaline solution (anion **38b**). The dependence of shift on pH follows the path of titration curves; it is possible to read off the pK value of the equilibrium from the point of inflection [2,6].

38a $\overset{1.52}{H_3C}-CH-CO_2^{\ominus}$ $\underset{+\,[H^{\oplus}]}{\overset{-\,[H^{\oplus}]}{\rightleftarrows}}$ $\overset{1.26}{H_3C}-CH-CO_2^{\ominus}$ **38b**

$\overset{\ominus}{N}H_3$ NH_2

pH = 6   pH = 12

^{13}C shifts respond to pH changes with even greater sensitivity; this is demonstrated by the values of pyridine (**39b**) and its cation (**39a**).

39a $\overset{142.5}{\underset{129.3}{}}\underset{148.4}{}$ $\underset{+\,[H^{\oplus}]}{\overset{-\,[H^{\oplus}]}{\rightleftarrows}}$ $\overset{149.9}{\underset{123.8}{}}\underset{136.0}{}$ **39a**

pH $\leqq$ 3   pH $\geqq$ 8

The effect of pH is rarely of use for pK measurement; it is more often of use in identifying the site of protonation/deprotonation when several basic or acidic sites are present. Knowing the incremental substituent effects Z [4-6] of amino and ammonium groups on benzene ring shifts in aniline and in the anilinium ion (**40**), one can decide which of the nitrogen atoms is protonated in procaine hydrochloride (problem 24).

40

^{13}C chemical shifts relative to benzene (δ_C = 128.5) as reference

2.6   Molecular dynamics (fluxionality)

2.6.1   Temperature-dependent NMR spectra

Figure 2.26 shows the ^1H NMR spectrum of *N,N*-dimethylacetamide (**41**) and its dependence on temperature. At 55 °C and below two resonances appear for the two *N*-methyl groups. Above 55 °C the signals become increasingly broad until they merge to form one broad signal at 80 °C. This temperature is referred to as the *coalescence temperature*, T_c. Above T_c the signal, which now belongs to both *N*-methyl groups, becomes increasingly sharp.

The temperature-dependent position and profile of the *N*-methyl signal result from amide canonical formulae of **41** shown in Fig. 2.26: the CN bond is a partial double bond; this hinders rotation of the *N,N*-dimethylamino group. One methyl group is now *cis* ($\delta_B = 3.0$) and the other is *trans* ($\delta_A = 2.9$) to the carboxamide oxygen. At low temperatures (55 °C), the *N*-methyl protons slowly exchange positions in the molecule (slow rotation, slow exchange). If energy is increased by heating (to above 90 °C), then the *N,N*-dimethylamino group rotates so that the *N*-methyl protons exchange their position with a high frequency (free rotation, rapid exchange), and one single, sharp *N*-methyl signal of double intensity appears with the average shift ($\delta_B + \delta_A)/2 = 2.95$.

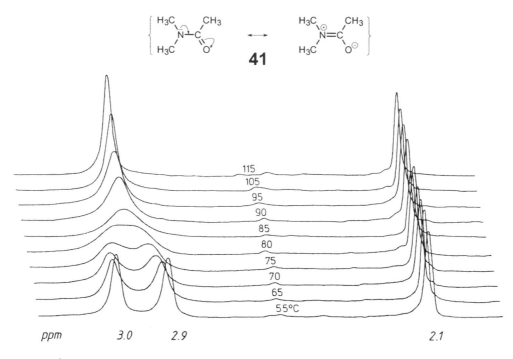

Figure 2.26. ^1H NMR spectra of *N,N*-dimethylacetamide (**41**) at the temperatures indicated [$(CD_3)_2SO$, 75 % v/v, 80 MHz]

The dimethylamino group rotation follows a first-order rate law; the exchanging methyl protons show no coupling and their singlet signals are of the same intensity. Under these conditions, equation 7 [2, 35-37] affords the rate constant k_r at the coalescence point T_c:

$$k_r = \pi(v_A - v_B)/\sqrt{2} = \pi \Delta v/\sqrt{2} \cong 2.22\,\Delta v \qquad (7)$$

where Δv is the full width at half-maximum of the signal at the coalesence point T_c; it corresponds to the difference in chemical shift ($v_B - v_A$) observed during slow exchange. In the case of dimethylacetamide (**41**) the difference in the chemical shift is *0.1 ppm* (Fig. 2.26), i.e. *8 Hz* (at 80 MHz). From equation 7 it can then be calculated that the *N*-methyl groups at the coalescence point (80 °C or 353 K) rotate with an exchange frequency of $k_r = 2.22 \times 8 = 17.8$ Hz. According to the Eyring equation 8, the exchange frequency k_r decreases exponentially with the free molar activation energy ΔG [35-37]:

$$k_r = \frac{kT_c}{h}\,e^{-\Delta G/RT_c} \qquad (8)$$

where R is the gas constant, k is the Boltzmann constant and h is the Planck constant. Equations 7 and 8 illustrate the value of temperature-dependent NMR for the investigation of molecular dynamics: following substitution of the fundamental constants, they give equation 9 for the free molar activation energy ΔG for first-order exchange processes:

$$\Delta G = 19.1\,T_c\,[\,10.32 + \log(T_c/k_r)\,] \times 10^{-3}\ kJ/mol \qquad (9)$$

Hence the activation energy barrier to dimethylamino group rotation in dimethylacetamide (**41**) is calculated from equation 9 with $k_r = 17.8 \text{ s}^{-1}$ at the coalescence point 353 K (Fig. 2.26):

$$\Delta G_{353} = 78.5 \text{ kJ / mol} \quad \text{or} \quad 18.7 \text{ kcal / mol}$$

Temperature-dependent (dynamic) NMR studies are suited to the study of processes with rate constants between 10^{-1} and 10^3 s^{-1} [3]. Some applications are shown in Table 2.13 and in problems 13 and 14.

Table 2.13. Selected applications of dynamic proton resonance [35-37]

			T_c (K)	ΔG_{T_c} (kJ / mol)
Rotation hindered by bulky substituents (*t*-butyl groups)			147	30
Inversion at amino-nitrogen (aziridines)			380	80
Ring inversion (cyclohexane)			193	25
Valence tautomerism (Cope systems, fluxionality)			298	3

2.6.2 ^{13}C Spin-lattice relaxation times

The spin-lattice relaxation time T_1 is the time constant with which an assembly of a particular nuclear spin in a sample becomes magnetised parallel to the magnetic field as it is introduced into it. The sample magnetisation M_0 is regenerated after every excitation with this time constant. For organic molecules the T_1 values of even differently bonded protons in solution are of the same order of magnitude (0.1 - 10 s). ^{13}C nuclei behave in a way which shows greater differentiation between nuclei and generally take more time: in molecules of varying size and in different chemical environments the spin-lattice relaxation times lie between a few milliseconds (macromolecules) and several minutes (quaternary C atoms in small molecules). With ^1H broadband decoupling only one T_1 value is recorded for each C atom (rather than n T_1 values as for all n components of a complex ^1H multiplet), and these ^{13}C spin-lattice relaxation times are useful parameters for probing molecular mobility in solution.

The technique for measurement which is most easily interpreted is the inversion-recovery method, in which the distribution of the nuclear spins among the energy levels is inverted by means of a suitable 180° radiofrequency pulse [2-6,17]. A negative signal is observed at first, which becomes increasingly positive with time (and hence also with increasing spin-lattice relaxation) and which finally approaches the equilibrium intensity asymptotically. Figures 2.27 and 2.28 show asymptotical increases in the signal amplitude due to ^{13}C spin-lattice relaxation up to the equilibrium value using two instructive examples. A simple analysis makes use of the 'zero intensity interval', τ_0, without consideration of standard deviations: after this time interval τ_0, the spin-lattice relaxation is precisely far enough advanced for the signal amplitude to pass through zero. Equation 10 then gives T_1 for each individual C atom.

$$T_1 = \tau_0 / \ln 2 \sim 1.45\ \tau_0 \qquad (10)$$

Thus, in the series of T_1 measurements of 2-octanol (**42**, Fig. 2.27) for the methyl group at the hydrophobic end of the molecule, the signal intensity passes through zero at $\tau_0 = 3.8$ s. From this, using equation 10, a spin-lattice relaxation time of $T_1 = 5.5$ s can be calculated. A complete relaxation of this methyl C atom requires about five times longer (more than 30 s) than is shown in the last experiment of the series (Fig. 2.27); T_1 itself is the time constant for an exponential increase, in other words, after T_1 the difference between the observed signal intensity and its final value is still 1/e of the final amplitude.

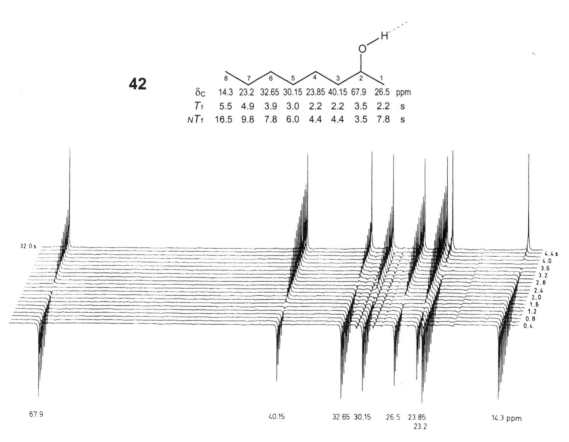

Figure 2.27. Sequence of measurements to determine the ^{13}C spin-lattice relaxation times of 2-octanol (**42**) [(CD$_3$)$_2$CO, 75% v/v, 25 °C, 20 MHz, inversion-recovery sequence, stacked plot]. The times at which the signals pass through zero, τ_0, have been used to calculate, by equation 10, the T_1 values shown above for the ^{13}C nuclei of 2-octanol

If smaller NOE enhancements are recorded for certain ^{13}C nuclei, then other mechanisms (e.g. spin-rotation) contribute to their spin-lattice relaxation [5,6].

The main contribution to the spin-lattice relaxation of ^{13}C nuclei which are connected to hydrogen is provided by the dipole-dipole interaction (DD mechanism, *dipolar relaxation*). For such ^{13}C nuclei a nuclear Overhauser enhancement of almost 2 will be observed during ^{1}H broadband decoupling according to:

$$\eta_c = \gamma_H / 2\gamma_C = 1.988 \qquad\qquad (11)$$

where γ_H and γ_C are the gyromagnetic constants of ^{1}H and ^{13}C.

Dipolar relaxation of ^{13}C nuclei originates from the protons (larger magnetic moment) in the same or in neighbouring molecules, which move with molecular motion (translation, vibration, rotation). This motion generates fluctuating local magnetic fields which affect the observed nucleus. If the frequency of a local magnetic field matches the Larmor frequency of the ^{13}C nucleus being observed (resonance condition), then this nucleus can undergo transition from the excited state to the ground state (relaxation) or the reverse (excitation). From this, it follows that the spin-lattice relaxation is linked to the mobility of the molecule or molecular fragment. If the average time taken between two reorientations of the molecule or fragment is defined as the effective correlation time τ_c, and if n H atoms are connected to the observed C, then the dipolar relaxation time $T_{1(DD)}$ is given by the correlation function:

$$T_{1(DD)}^{-1} = \text{constant} \times n\,\tau_c \qquad\qquad (12)$$

Accordingly, the relaxation time of a C atom will increase the fewer hydrogen atoms it bonds to and the faster the motion of the molecule or molecular fragment in which it is located. From this, it can be deduced that the spin-lattice relaxation time of ^{13}C nuclei provides information concerning four molecular characteristics:

■ **Molecular size:** smaller molecules move more quickly than larger ones; as a result, C atoms in small molecules relax more slowly than those in large molecules. The C atoms in the more mobile cyclohexanes ($T_1 = 19\text{-}20$ s) take longer than those in the more sluggish cyclodecane ($T_1 = 4\text{-}5$ s) [5,6].

■ **The number of bonded H atoms:** if all parts within a molecule move at the same rate (the same τ_c for all C atoms), the relaxation times T_1 decrease from CH via CH_2 to CH_3 in the ratio given by:

$$T_1(CH) : T_1(CH_2) : T_1(CH_3) = 6:3:2 \qquad\qquad (13)$$

Since methyl groups also rotate freely in otherwise rigid molecules, they follow the ratio shown in equation 13 only in the case of considerable steric hindrance [6]. In contrast, the T_1 values of ^{13}C nuclei of CH and CH_2 groups follow the ratio 2 : 1 even in large, rigid molecules. Typical examples are steroids such as cholesteryl chloride (**43**), in which the CH_2 groups of the ring relax at approximately double the rate (0.2 - 0.3 s) of CH carbon atoms (0.5 s). Contrary to the prediction made by equation 13, freely rotating methyl groups require considerably longer (1.5 s) for spin-lattice relaxation.

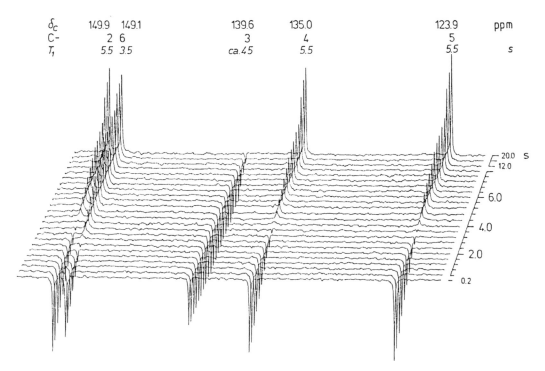

- **Segmental mobility:** if one examines the T_1 series of 2-octanol (**42**, Fig. 2.27) calculated according to equation 10, it becomes apparent that the mobility parameters nT_1 increase steadily from C-2 to C-8. As a result of hydrogen bonding, the molecule close to the OH groups is almost rigid (nT_1 between 3.5 and 4.4 s). With increasing distance from the anchoring effect of the OH group the mobility increases; the spin-lattice relaxation time becomes correspondingly longer. The nT_1 values of the two methyl groups also reflect the proximity to (7.8 s) and distance from (16.5 s) the hydrogen bond as a 'braking' device.

Figure 2.28. Sequence of measurements to determine the spin-lattice relaxation times of the ^{13}C nuclei of the pyridine ring in L-nicotine (**46**) [$(CD_3)_2CO$, 75 % v/v, 25 °C, inversion-recovery sequence, 20 MHz]. The times at which signals pass through zero have been used to calculate, by equation 10, the T_1 values for the pyridine C atoms in L-nicotine

- **Anisotropy of molecular motion:** monosubstituted benzene rings, e.g. phenyl benzoate (**44**), show a very typical characteristic: in the *para* position to the substituents the CH nuclei relax considerably more rapidly than in the *ortho* and *meta* positions. The reason for this is the anisotropy of the molecular motion: the benzene rings rotate more easily around an axis which passes through the substituents and the *para* position, because this requires them to push aside the least number of neighbouring molecules. This rotation, which affects only the *o*- and *m*-C*H* units, is too rapid for an effective spin-lattice relaxation of the *o*- and *m*-C atoms. More efficient with respect to relaxation are the frequencies of molecular rotations perpendicular to the preferred axis, and these affect the C*H* bond in *p*-position. If the phenyl rotation is impeded by bulky substituents, e.g. in 2,2′,6,6′-tetramethylbiphenyl (**45**), then the T_1 values of the C*H* atoms can be even less easily distinguished in the *meta* and *para* positions (3.0 and 2.7 s, respectively). Particularly large spin-lattice relaxation times obtained for non-protonated carbon nuclei of **45** and **46** arise from less efficient, non-dipolar relaxation mechanisms.

44   **45**   **46**

Figure 2.28 shows the anisotropy of the rotation of the pyridine ring in nicotine (**46**). The main axis passes through C-3 and C-6; C-6 relaxes correspondingly more rapidly (3.5 s) than the three other C*H* atoms (5.5 s) of the pyridine ring in nicotine, as can be seen from the times at which the appropriate signals pass through zero.

2.7 Summary

Table 2.14 summarizes the steps by which molecular structures can be determined using the NMR methods discussed thus far to determine the skeleton structure, relative configuration and conformation of a specific compound.

In the case of completely unknown compounds, the molecular formula is a useful source of additional information; it can be determined using small amounts of substance (a few micrograms) by high-resolution mass spectrometric determination of the accurate molecular mass. It provides information concerning the double-bond equivalents (the 'degree of unsaturation' – the number of multiple bonds and rings). For the commonest heteroatoms in organic molecules (nitrogen, oxygen, sulphur, halogen), the number of double-bond equivalents can be derived from the molecular formula by assuming that oxygen and sulphur may be omitted and require no replacement atom, halogen may be replaced by hydrogen and nitrogen may be replaced by C*H*. The resulting empirical formula C_nH_x is then compared with the empirical formula of an alkane with *n* C atoms, C_nH_{2n+2}; the number of double-bond equivalents is equal to half the hydrogen deficit,

$$(2n + 2 - x) \,/\, 2 \,.$$

From C_8H_9NO (problem 4), for example, the empirical formula C_9H_{10} is derived and compared with the alkane formula C_9H_{20}; a hydrogen deficit of ten and thus of five double-bond equivalents is deduced. If the NMR spectra have too few signals in the shift range appropriate for multiple bonds, then the double-bond equivalents indicate rings (see, for example, α-pinene, Fig. 2.4).

If the amount of the sample is sufficient, then the carbon skeleton is best traced out from the two-dimensional INADEQUATE experiment. If the absolute configuration of particular C atoms is needed, the empirical applications of diastereotopism and chiral shift reagents are useful (Section 2.4). Anisotropic and ring current effects supply information about conformation and aromaticity (Section 2.5), and pH effects can indicate the site of protonation (problem 24). Temperature-dependent NMR spectra and ^{13}C spin-lattice relaxation times (Section 2.6) provide insight into molecular dynamics (problems 13 and 14).

Table 2.14. Suggested tactics for solving structures using NMR

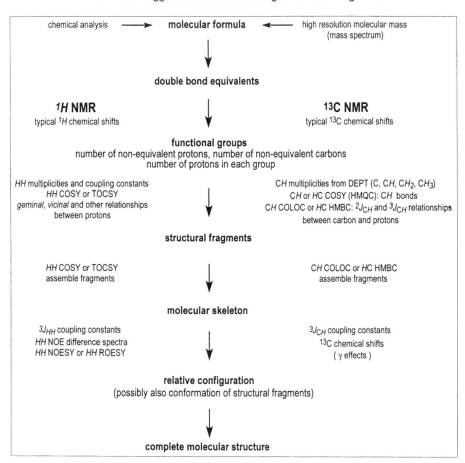

3 PROBLEMS

In the following 55 problems, the chemical shift value (ppm) is given in the scale below the spectra and the coupling constant (Hz) is written immediately above or below the appropriate multiplet. Proton NMR data are italicised throughout in order to distinguish them from the parameters of other nuclei (^{13}C, ^{15}N).

Problem 1

^{1}H NMR spectrum **1** was obtained from dimethyl cyclopropanedicarboxylate. Is it a *cis* or a *trans* isomer?

Conditions: CDCl$_3$, 25 °C, 400 MHz.

1

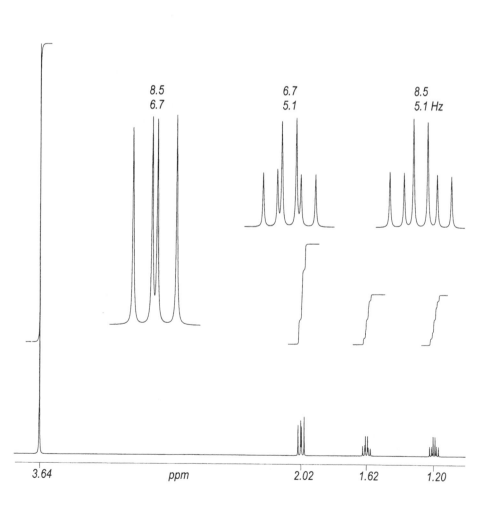

Problem 2

From which compound of formula $C_5H_8O_2$ was ^1H NMR spectrum **2** obtained?
Conditions: CDCl$_3$, 25 °C, 90 MHz.

2

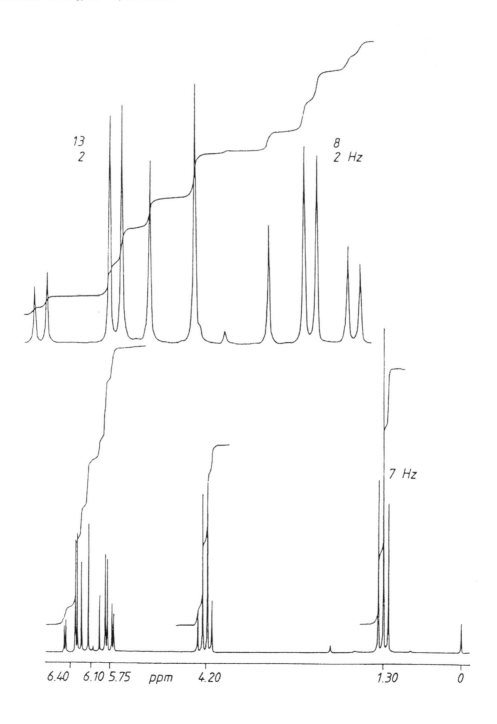

13
2

8
2 Hz

7 Hz

6.40   6.10 5.75   ppm   4.20   1.30   0

Problem 3

Which stereoisomer of the compound C_5H_6O is present given spectrum **3**?
Conditions: CDCl$_3$, 25 °C, 90 MHz.

3

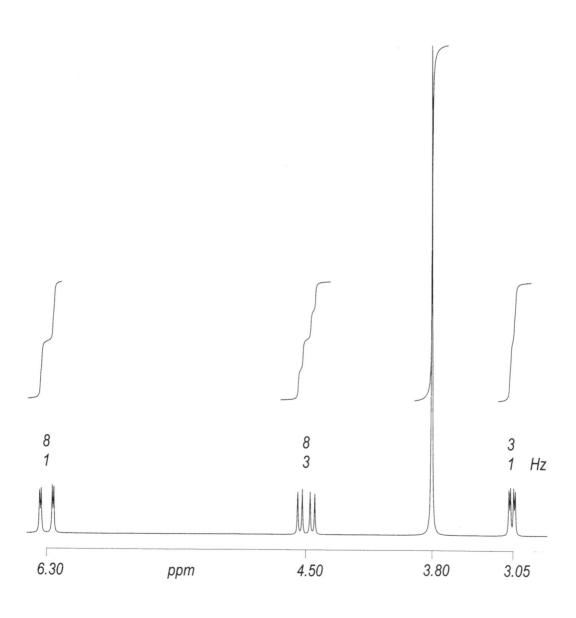

Problem 4

Which stereoisomer of the compound C_8H_9NO can be identified from ^1H NMR spectrum **4**?
Conditions: $CDCl_3$, 25 °C, 90 MHz.

4

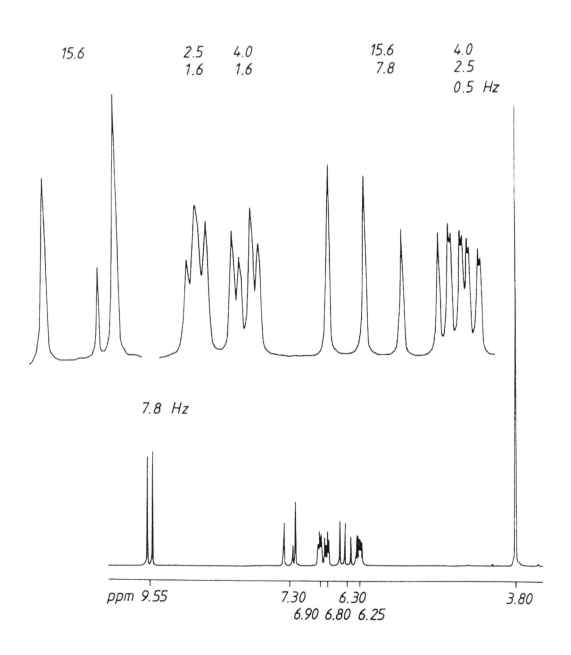

Problem 5

The reaction of 2,2'-bipyrrole with orthoformic acid triethyl ester in the presence of phosphoryl chloride ($POCl_3$) produced a compound which gave the ^1H NMR spectrum **5**. Which compound has been prepared?

Conditions: $CDCl_3$, 25 °C, 400 MHz.

Two broad D_2O-exchangeable signals at $\delta_H = 11.6$ (one proton) and 12.4 (two protons) are not shown.

5

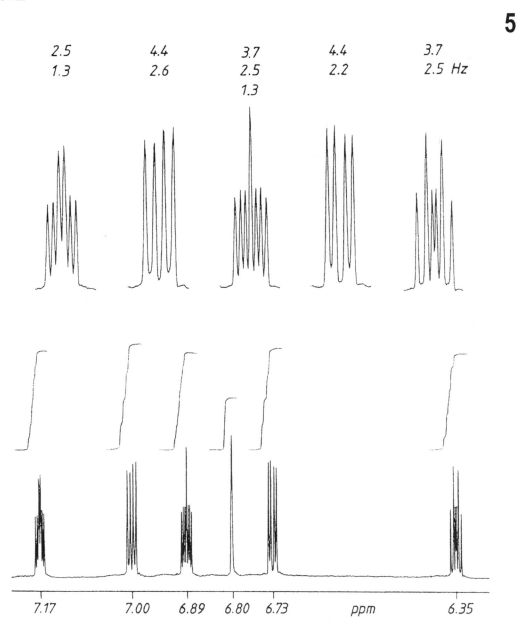

2.5	4.4	3.7	4.4	3.7
1.3	2.6	2.5	2.2	2.5 Hz
		1.3		

7.17   7.00   6.89   6.80   6.73   ppm   6.35

Problem 6

From which compound C_7H_7NO was ^1H NMR spectrum **6** obtained ?
Conditions: $CDCl_3$, 25 °C, 90 MHz.

6

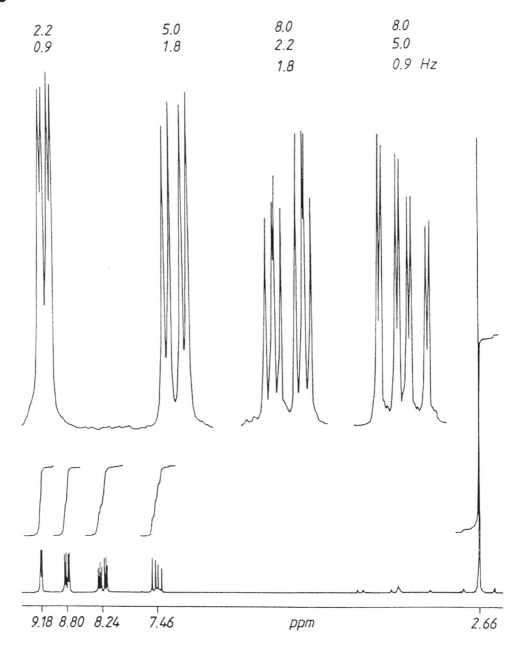

Problem 7

From which compound $C_{16}H_{10}O_2$ was ^1H NMR spectrum **7** obtained ?
Conditions: $CDCl_3$, 25 °C, 400 MHz.

7

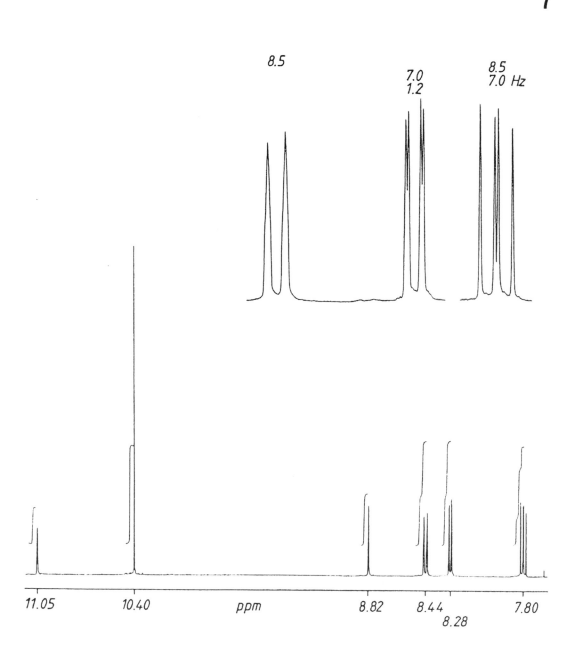

Problem 8

One functional group and three substructures given by the ^1H NMR spectrum **8** make up the molecular structure including the relative configuration of the sample compound.

Conditions: CDCl$_3$, 25 °C, 400 MHz.

8

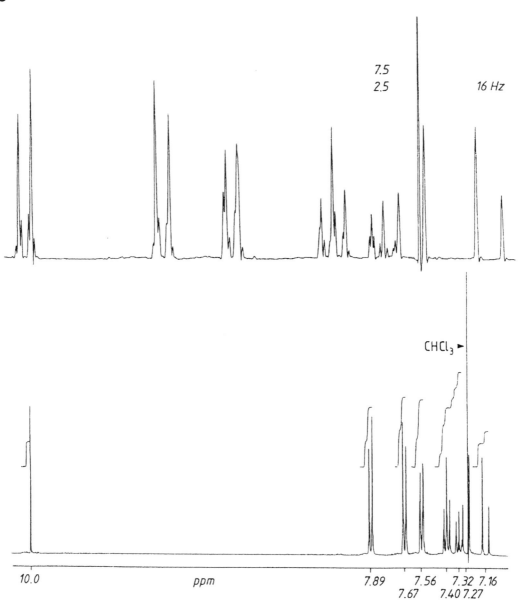

7.5
2.5
16 Hz

CHCl$_3$ ►

10.0   ppm   7.89   7.56   7.32 7.16
                                                                                        7.67   7.40 7.27

Problem 9

Which substituted isoflavone can be identified from ^1H NMR spectrum **9**?
Conditions: CDCl₃, 25 °C, 200 MHz.

9

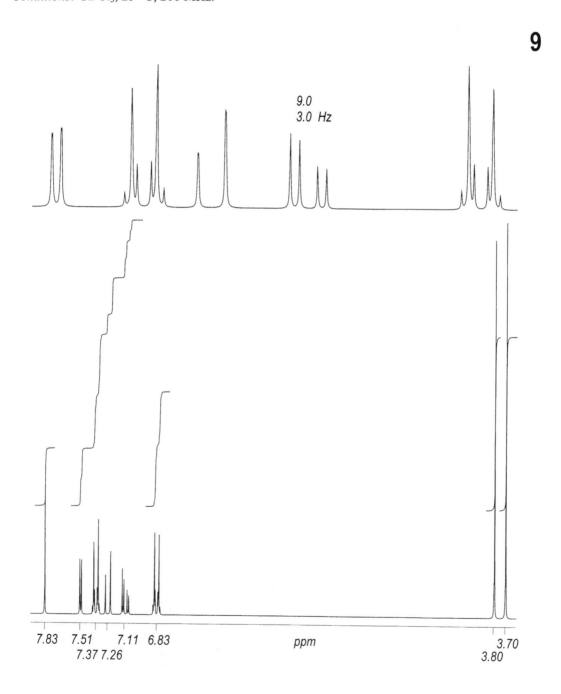

9.0
3.0 Hz

7.83   7.51   7.11 6.83   *ppm*   3.70
7.37 7.26   3.80

Problem 10

A natural substance of elemental composition $C_{15}H_{14}O_6$ was isolated from the plant *Centaurea chilensis* (Compositae). What is the structure and relative configuration of the substance given its ^1H NMR spectrum **10** with (top) and without (bottom) deuterium exchange **10**?

Conditions: CDCl₃, 25 °C, 400 MHz.

10

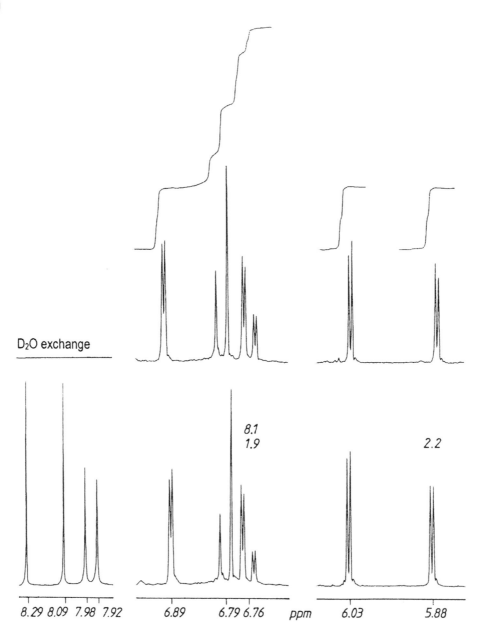

D₂O exchange

8.1
1.9

2.2

8.29 8.09 7.98 7.92   6.89   6.79 6.76   *ppm*   6.03   5.88

Problem 10, continued

10

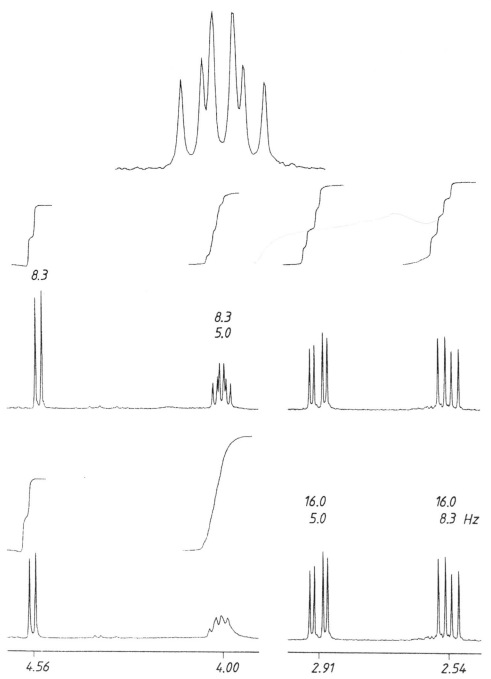

8.3

8.3
5.0

16.0   16.0
5.0   8.3 Hz

4.56   4.00   2.91   2.54

Problem 11

Characterisation of the antibiotic monordene (also referred to as radicicol) with the elemental composition $C_{18}H_{17}O_6Cl$ isolated from *Monosporium bonorden* gave the macrolide structure *1*. The relative configuration of the *H* atoms on the two conjugated double bonds (6,7-*cis*, 8,9-*trans*-) could be deduced from the 60 MHz ^1H NMR spectrum [38]. The relative configuration of the C atoms 2-5, which encompass the oxirane ring as a partial structure, has yet to be established.

The reference compound methyloxirane gives the ^1H NMR spectrum **11a** shown with expanded multiplets. What information regarding its relative configuration can be deduced from the expanded ^1H multiplets of monordene displayed in **11b**?

Conditions: $(CD_3)_2CO$, 25 °C, 200 MHz.

11a

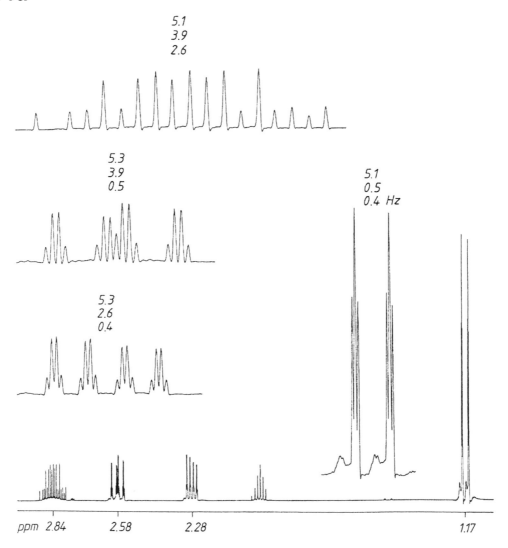

ppm 2.84   2.58   2.28   1.17

Problem 11, continued

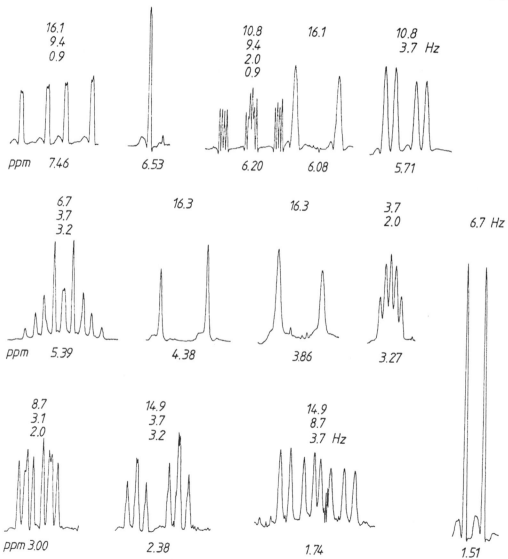

11b

Problem 12

From the *HH* COSY contour plot **12a** it can be established which cycloadduct has been produced from 1-(*N,N*-dimethylamino)-2-methylbuta-1,3-diene and *trans*-β-nitrostyrene. The $^3J_{HH}$ coupling constant in the one-dimensional ^1H NMR spectrum **12b** can be used to deduce the relative configuration of the adduct.

Conditions: CDCl$_3$, 25 °C, 400 MHz.

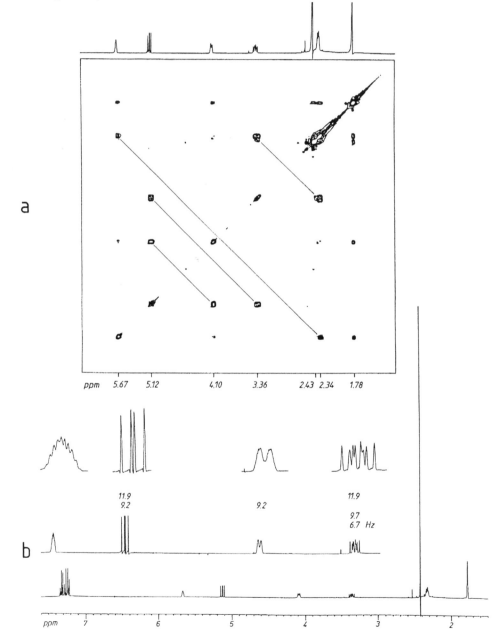

Problem 13

^1H NMR spectra **13** were recorded for 3-(*N,N*-dimethylamino)acrolein at the temperatures given. What can be said about the structure of the compound and what thermodynamic data can be derived from these spectra?

Conditions: CDC1$_3$, 50 % v/v, 250 MHz.

13

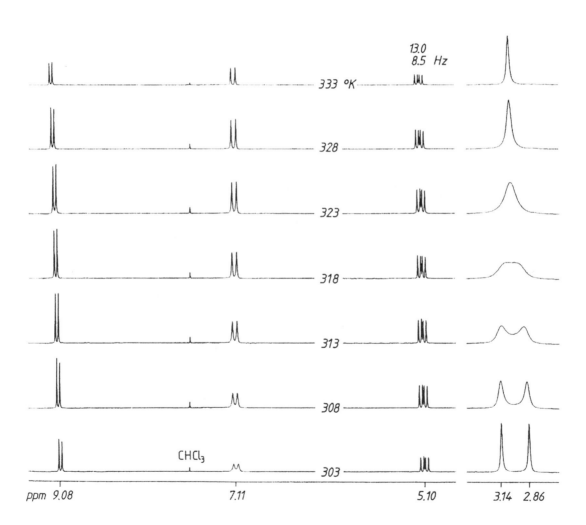

Problem 14

^{13}C NMR spectra **14** were recorded of *cis*-1,2-dimethylcyclohexane at the temperatures given; the DEPT experiment at 223 K was also recorded in order to distinguish the C*H* multiplicities (C*H* and C*H₃* positive, C*H₂* negative). Which assignments of resonances and what thermodynamic data can be deduced from these spectra ?

Conditions: (CD₃)₂CO, 95 % v/v, 100 MHz, ^{1}*H* broadband decoupled.

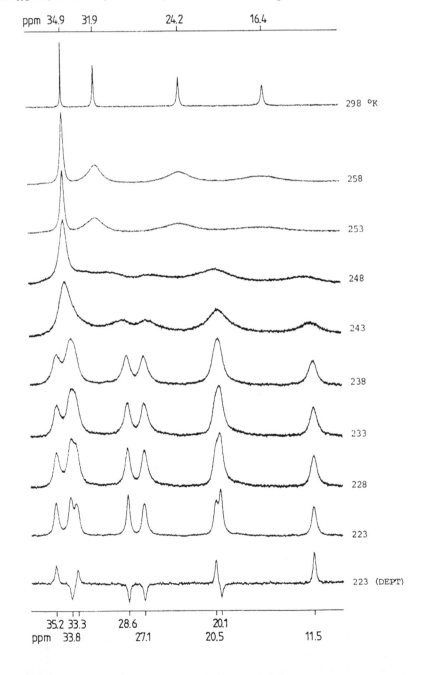

Problem 15

No further information is required to identify this compound from its ^{13}C NMR spectra **15**.

Conditions: CDCl₃, 25 °C, 20 MHz. (**a**) Proton broadband decoupled spectrum; (**b**) NOE enhanced coupled spectrum (gated decoupling); (**c**) expanded section of (**b**).

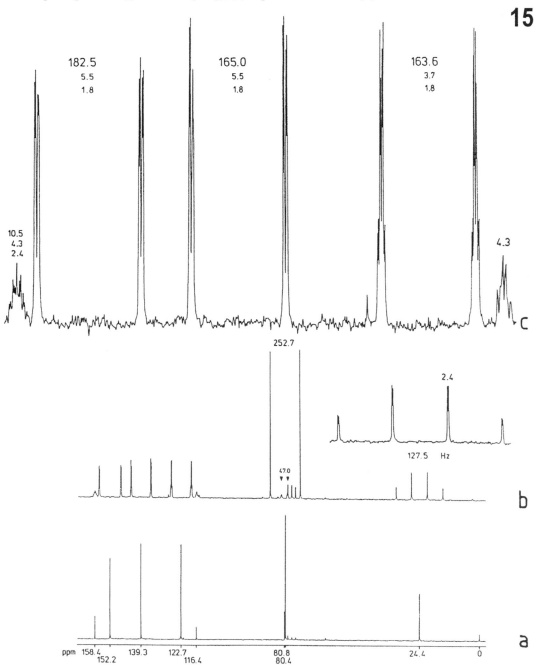

Problem 16

In hexadeuteriodimethyl sulphoxide the compound which is labelled as 3-methyl-pyrazolone gives
^{13}C NMR spectra **16**. In what form is this compound present in this solution?

Conditions: $(CD_3)_2SO$, 25 °C, 20 MHz. (**a**) ^1H broadband decoupled spectrum; (**b**) NOE enhanced
coupled spectrum (gated decoupling); (**c**) expanded sections of (**b**).

16

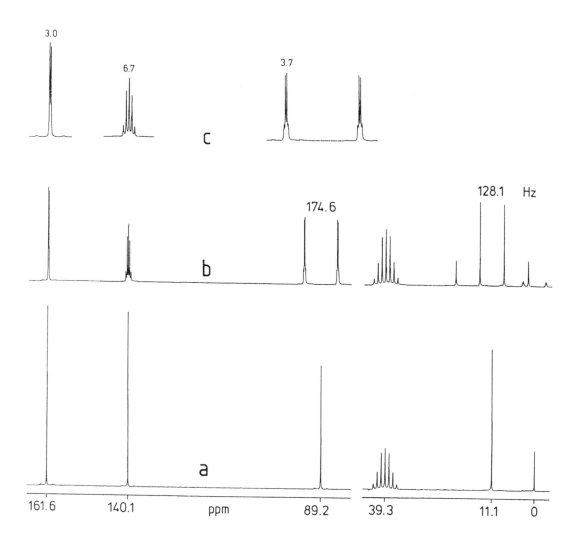

Problem 17

The compound $C_8H_8O_2$ gave the ^{13}C NMR spectra **17**. What is its structure?

Conditions: $CDCl_3 : (CD_3)_2CO$ (1 : 1), 25 °C, 20 MHz. (**a**) ^1H broadband decoupled spectrum; (**b**) NOE enhanced coupled spectrum (gated decoupling); (**c**) expanded section of (**b**), δ_C = 118.2-136.5.

17

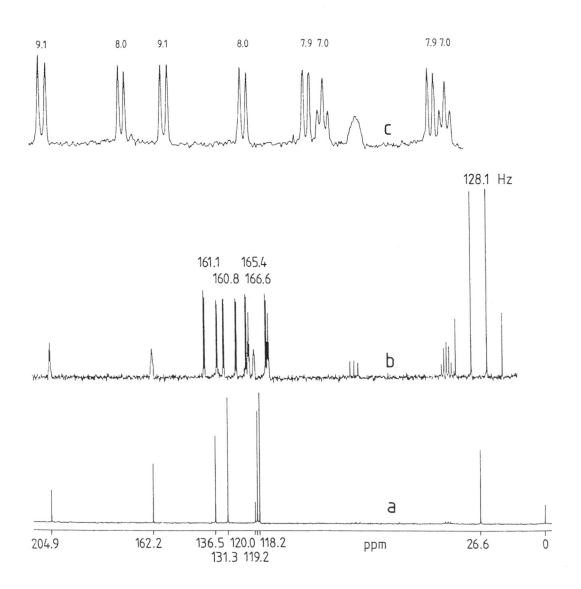

Problem 18

1,3,5-Trinitrobenzene reacts with dry acetone in the presence of potassium methoxide to give a crystalline violet compound $C_9H_8N_3O_7K$. Deduce its identity from the ^{13}C NMR spectra **18**.

Conditions: $(CD_3)_2SO$, 25 °C, 22.63 MHz. (**a**) ^1H broadband decoupled spectrum; (**b**) without decoupling; (**c**) expanded section of (**b**).

18

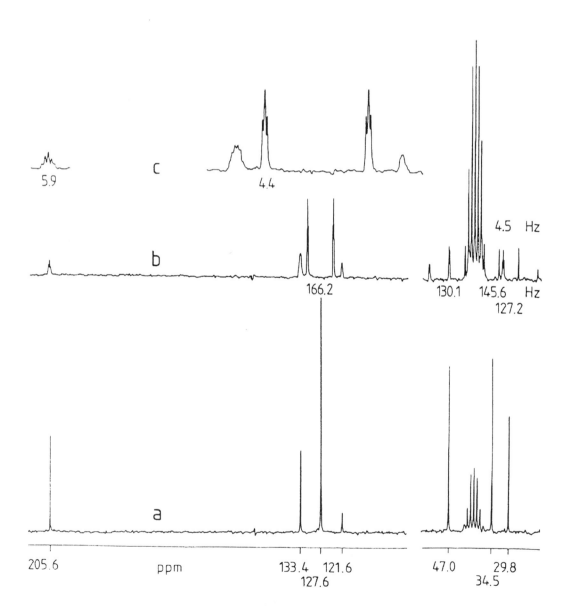

Problem 19

3-[4-(*N*,*N*-Dimethylamino)phenyl]-2-ethylpropenal (*3*) was produced by reaction of *N*,*N*-dimethyl-aniline (*1*) with 2-ethyl-3-ethoxyacrolein (*2*) in the presence of phosphorus oxytrichloride.

Since the olefinic CC double bond is trisubstituted, the relative configuration cannot be deter-mined on the basis of the *cis* and *trans* couplings of *vicinal* alkene protons in the ^1H NMR spec-trum. What is the relative configuration given the ^{13}C NMR spectra **19**?

Conditions: CDCl₃, 25 °C, 20 MHz. (**a**) ^1H broadband decoupled spectrum; (**b**) expanded sp³ shift range; (**c**) expanded sp² shift range; (**b**) and (**c**) each with the ^1H broadband decoupled spectrum below and NOE enhanced coupled spectrum above.

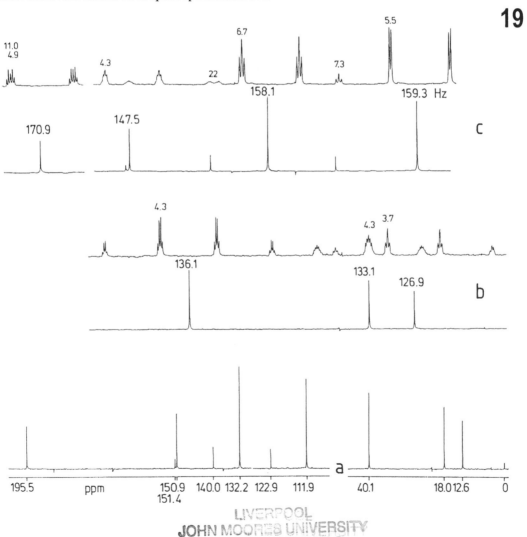

Problem 20

2-Trimethylsilyloxy-β-nitrostyrene was the target of Knoevenagel condensation of 2-trimethyl-siloxybenzaldehyde with nitromethane in the presence of *n*-butylamine as base. ^{13}C NMR spectra **20** were obtained from the product of the reaction. What has happened?

Conditions: CDCl$_3$, 25 °C, 20 MHz. (**a**) sp^3 shift range; (**b**) sp^2 shift range, in each case with the ^{1}H broadband decoupled spectrum below and the NOE enhanced coupled spectrum (gated decoupling).

20

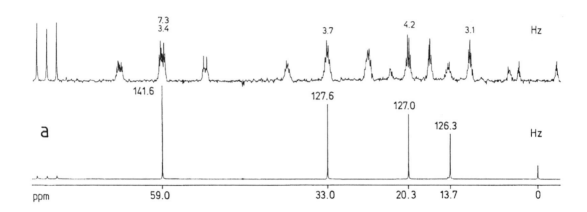

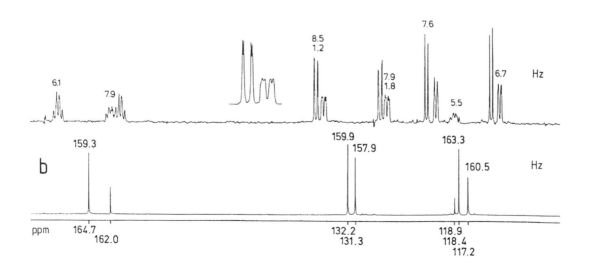

Problem 21

From which compound were the INADEQUATE contour plot and ^{13}C NMR spectra **21** obtained?

Conditions: $(CD_3)_2CO$, 95 % v/v, 25 °C, 100 MHz. (**a**) Symmetrised INADEQUATE contour plot with ^{13}C NMR spectra; (**b**) ^1H broadband decoupled spectrum; (**c**) NOE enhanced coupled spectrum (gated decoupling); (**d**) expansion of multiplets of (**c**).

21

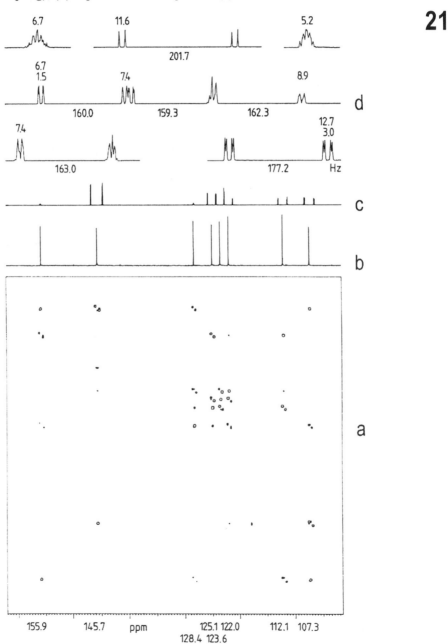

Problem 22

The hydrolysis of 3-ethoxy-4-ethylbicyclo[4.1.0]hept-4-en-7-one propylene acetal (*1*) with aqueous acetic acid in tetrahydrofuran gives an oil with the molecular formula $C_{12}H_{18}O_3$, from which the INADEQUATE contour plot **22** and DEPT spectra were obtained. What is the compound?

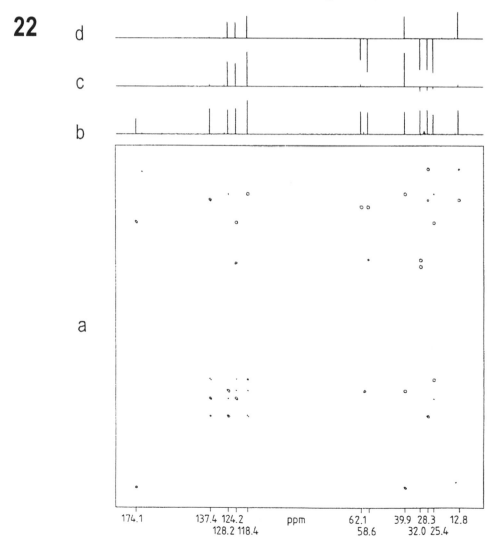

Conditions: $(CD_3)_2CO$, 95 % v/v, 25 °C, 100 MHz. (**a**) Symmetrised INADEQUATE contour plot; (**b-d**) ^{13}C NMR spectra, (**b**) ^1H broadband decoupled spectrum, (**c**) DEPT CH subspectrum, (**d**) DEPT subspectrum, CH and CH_3 positive and CH_2 negative.

Problem 23

From which compound $C_6H_{10}O$ were the NMR spectra **23** recorded?

Conditions: $(CD_3)_2CO$, 25 °C, 200 MHz (^1H), 50 MHz (^{13}C). (**a**) ^1H NMR spectrum with expanded sections; (**b,c**) ^{13}C NMR partial spectra, each with proton broadband decoupled spectrum below and NOE enhanced coupled spectrum above with expanded multiplets at $\delta_C = 76.6$ and 83.0.

23

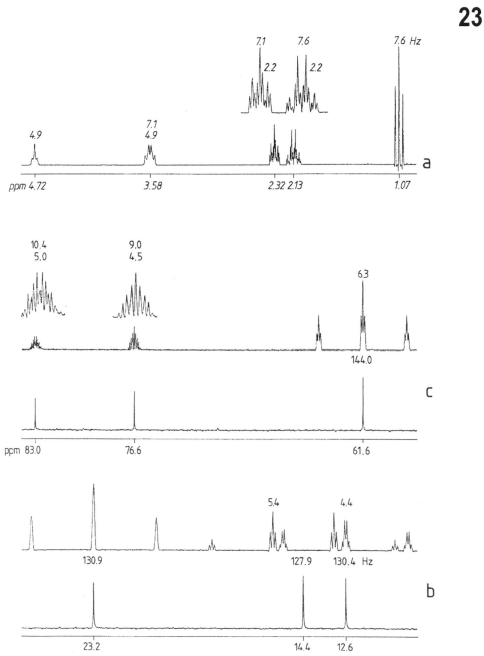

Problem 24

Procaine hydrochloride gives the ^1H and ^{13}C NMR spectra **24**. Which amino group is protonated?
Conditions: $CDCl_3 : (CD_3)_2SO$ (3 : 1), 25 °C, 20 MHz (^{13}C), 200 MHz (^1H). (**a**) ^1H NMR spectrum
with expanded multiplets; (**b-d**) ^{13}C NMR spectra; (**b**) ^1H broadband decoupled spectrum; (**c**) NOE
enhanced coupled spectrum (gated decoupling); (**d**) expansion of multiplets (δ_C = 113.1 - 153.7).

24

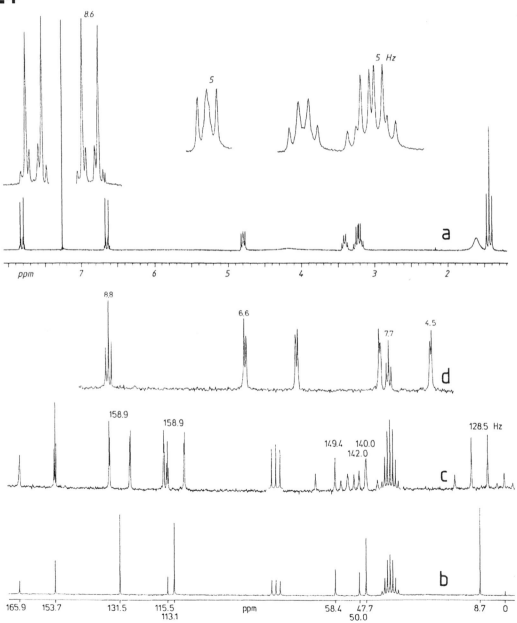

Problem 25

Using which compound $C_{10}H_{10}O_4$ were the NMR spectra **25** recorded?

Conditions: $(CD_3)_2CO$, 25 °C, 400 MHz (^1H), 100 MHz (^{13}C). (**a**) ^1H NMR spectrum with expanded sections; (**b,c**) ^{13}C NMR spectra, (**b**) proton broadband decoupled, (**c**) NOE enhanced coupled spectrum with expanded sections.

25

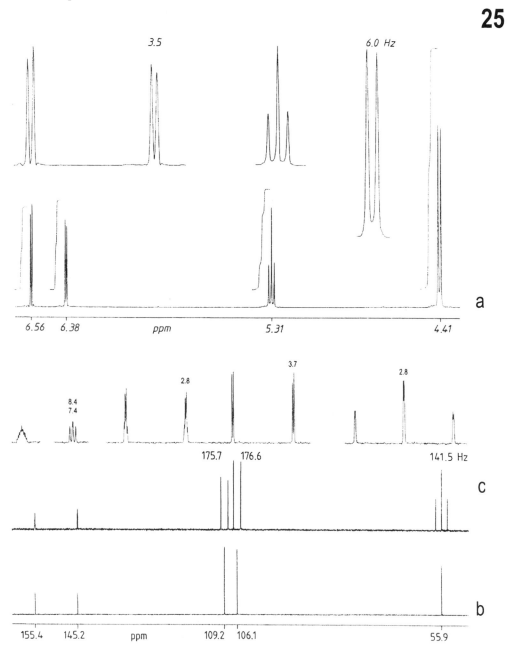

Problem 26

An alkaloid of elemental composition $C_{11}H_{11}NO_3$ was isolated from the plant *Dactylicapnos to-rulosa* (Fumariaceae). What is the structure of it given its NMR spectra **26**?

Conditions: $CDCl_3$, 25 °C, 400 MHz (^1H), 100 MHz (^{13}C). (**a**) ^1H NMR spectrum with expanded section; (**b,c**) ^{13}C NMR spectra, proton broadband decoupled (bottom) and NOE enhanced coupled with expanded sections (top), (**b**) from $\delta_C = 27.6\text{-}47.8$, (**c**) from $\delta_C = 101.3\text{-}164.2$.

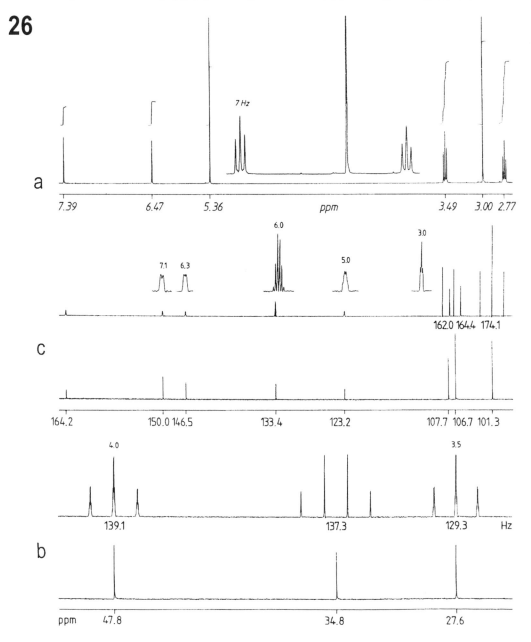

Problem 27

3,4-Dihydro-2*H*-pyran-5-carbaldehyde (*1*) was treated with sarcosine ethylester (*2*) in the presence of sodium ethoxide. What is the structure of the crystalline product $C_{11}H_{16}NO_3$ given its NMR spectra **27**?

Conditions: CDCl$_3$, 25 °C, 100 MHz (^{13}C), 400 MHz (^{1}H). (**a**) ^{1}H NMR spectrum with expanded multiplets; (**b-e**) ^{13}C NMR partial spectra; (**b**) sp^3 shift range; (**c**) sp^2 shift range, each with ^{1}H broadband decoupled spectrum (bottom) and the NOE enhanced coupled spectrum (gated decoupling, top) with expansions of multiplets (**d**) (δ_C =59.5 - 61.7) and (**e**) (δ_C=117.0 - 127.5).

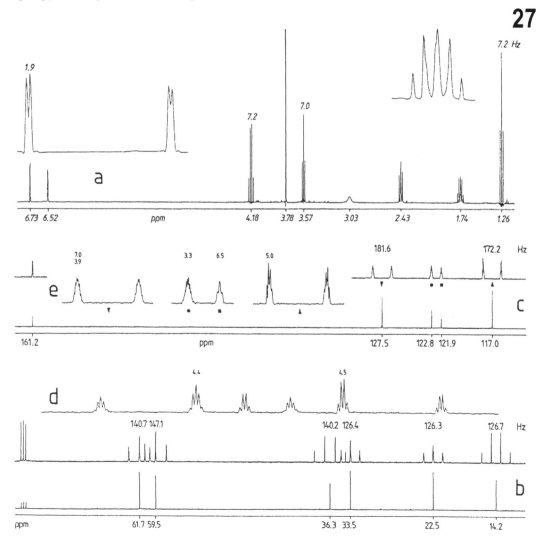

Problem 28

1-Ethoxy-2-propylbuta-1,3-diene and *p*-tolyl sulphonyl cyanide react to give a crystalline product. What is this product given its NMR spectra set **28**?

Conditions: CDCl$_3$, 25 °C, 100 MHz (^{13}C), 400 MHz (^1H). (**a-e**) ^{13}C NMR spectra; (**a,b**) ^1H broadband decoupled spectra; (**c,d**) NOE enhanced coupled spectra (gated decoupling) with expansion (**e**) of the multiplets in the sp^2 shift range; (**f**) ^1H NMR spectrum with expanded multiplets.

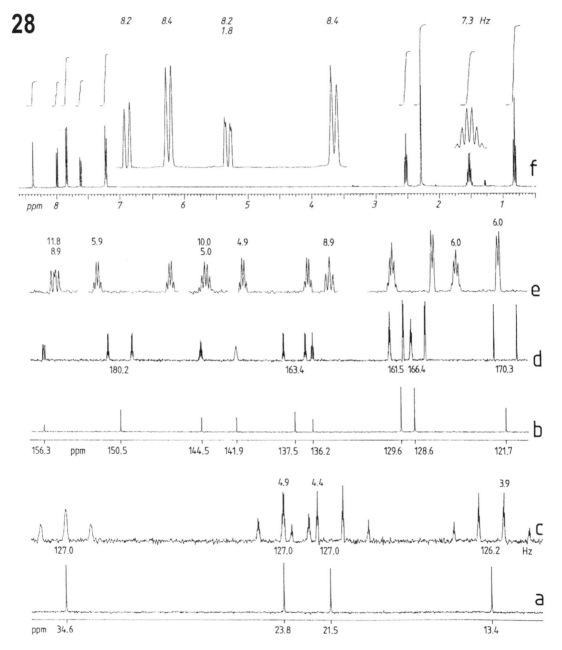

Problem 29

Which compound of formula $C_{11}H_{12}O_2$ gives the NMR spectra set **29**?

Conditions: CDCl₃, 25°C, 200 MHz (¹H), 50 MHz (¹³C). (**a**) ¹H NMR spectrum with expansion (**b**) and NOE difference spectra (**c,d**), with decoupling at δ_H = *2.56* (**c**) and *2.87* (**d**); (**e-g**) ¹³C NMR spectra; (**e**) ¹H broadband decoupled spectrum; (**f**) NOE enhanced coupled spectrum (gated decoupling) with expansions (**g**) (δ_C = 23.6, 113.3, 113.8, 127.0, 147.8, 164.6 and 197.8); (**h,** next page) section of the C*H* COSY diagram.

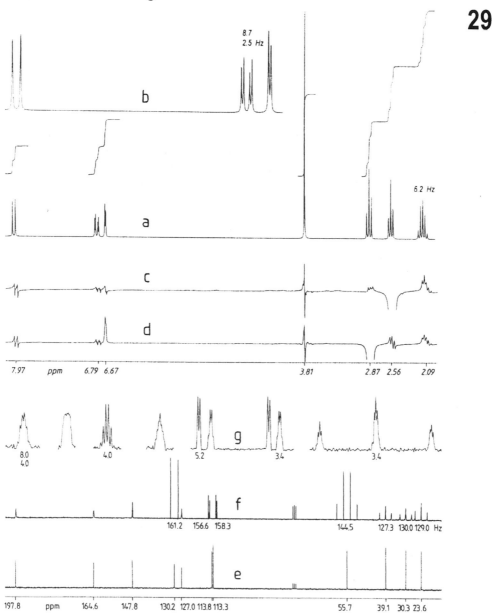

Problem 29, continued

29h

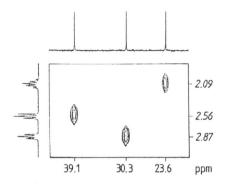

Problem 30

5-Amino-l,2,4-triazole undergoes a cyclocondensation with 3-ethoxyacrolein (*1*) to form 1,2,4-triazolo[1,5-*a*]pyrimidine (*3*) or its [4,3-*a*] isomer (*5*), according to whether it reacts as *1H* or *4H* tautomer *2* or *4*. Moreover, the pyrimidines *3* and *5* can interconvert by a Dimroth rearrangement. Since the ^1H NMR spectrum **30a** does not enable a clear distinction to be made (*AMX* systems for the pyrimidine protons in both isomers), the ^{13}C and ^{15}N NMR spectra **30b-d** were obtained. What is the compound?

30a

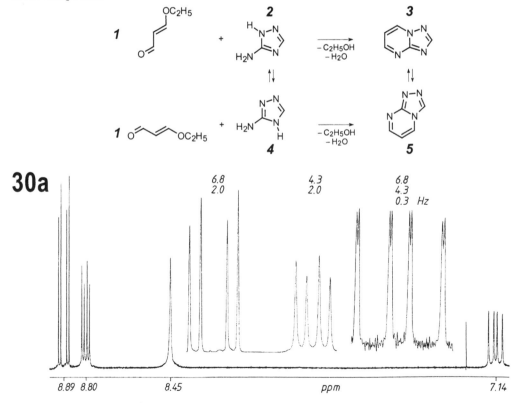

Problem 30, continued

Conditions: CDCl₃ (**a-c**), (CD₃)₂SO (**d**), 25 °C, 200 MHz (*¹H*), 20 MHz (*¹³C*), 40.55 MHz (*¹⁵N*). (**a**) *¹H* NMR spectrum; (**b,c**) *¹³C* NMR spectra; (**b**) *¹H* broadband decoupled spectrum; (**c**) NOE enhanced coupled spectrum (gated decoupling); (**d**) *¹⁵N* NMR spectrum, without decoupling, with expanded multiplets, *¹⁵N* shifts calibrated relative to ammonia as reference [7, 8].

30

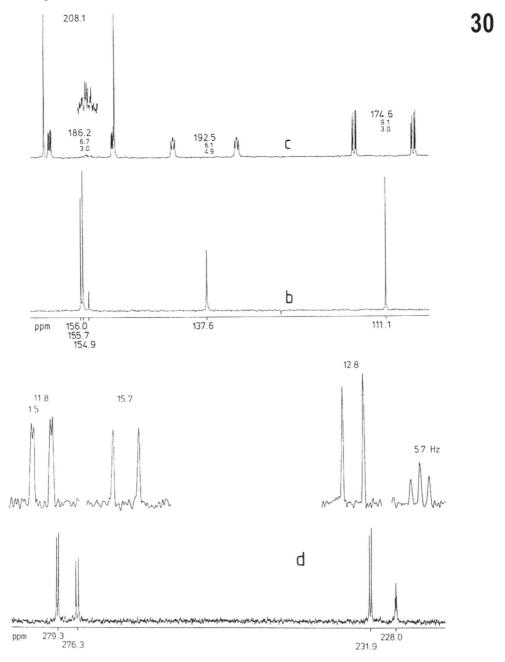

Problem 31

NMR spectra **31** were obtained from 6-butyltetrazolo[1,5-*a*]pyrimidine (*1*). What form does the heterocycle take?

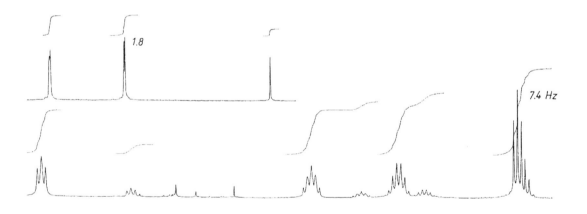

1

Conditions: (CD$_3$)$_2$CO, 25 °C, 400 MHz (^1H), 100 MHz (^{13}C), 40.55 MHz (^{15}N). (**a**) ^1H NMR spectrum with expanded partial spectra and integrals; (**b, c**) ^{13}C NMR spectra, in each case showing proton broadband decoupled spectrum below and gated decoupled spectrum above, (**b**) aliphatic resonances and (**c**) heteroaromatic resonances; (**d**) ^{15}N NMR spectrum, coupled, with expanded sections and integrals.

31a

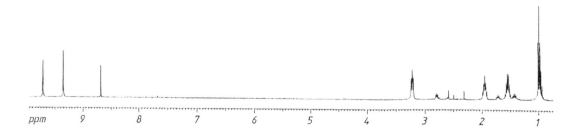

Problem 31, continued

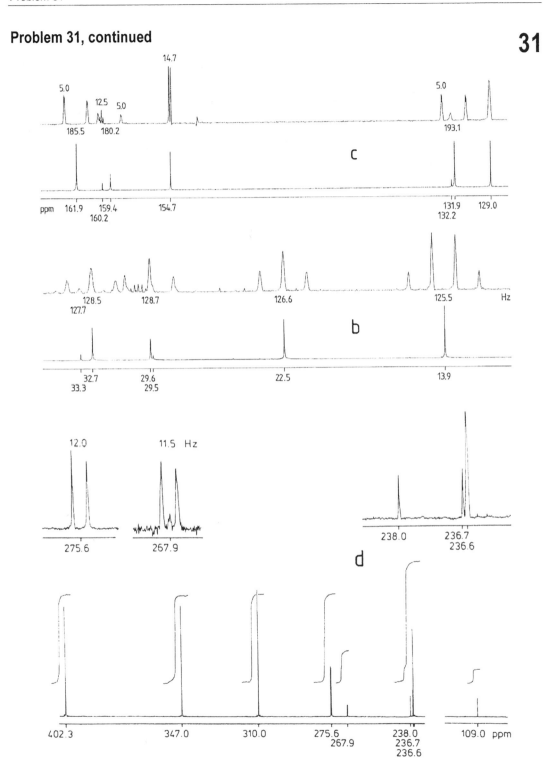

Problem 32

Phthalaldehydic acid (*o*-formylbenzoic acid) gives the NMR spectra set **32**. In what form does the compound actually exist?

Conditions: CDCl$_3$: (CD$_3$)$_2$SO (9:1), 25 °C, 80 and 400 MHz (^1H), 20 and 100 MHz (^{13}C). (**a**) ^1H NMR spectrum with expanded section (**b**) before and after D$_2$O exchange; (**c,d**) ^{13}C NMR spectra; (**c**) ^1H broadband decoupled spectrum; (**d**) NOE enhanced coupled spectrum (gated decoupling); (**e**) *CH* COSY diagram (100/400 MHz); (**f**) *HH* COSY diagram (400 MHz). The ordinate scales in (**e**) and (**f**) are the same.

32

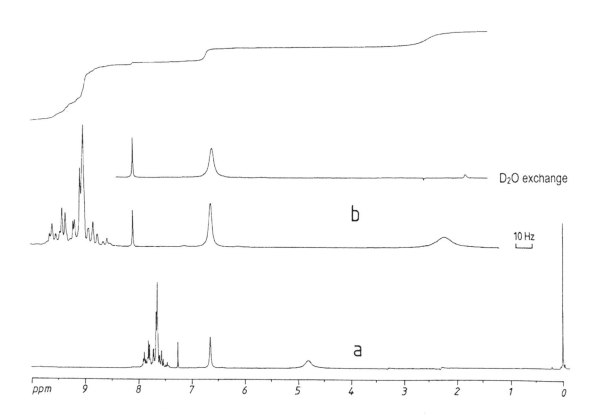

Problem 32, continued

32

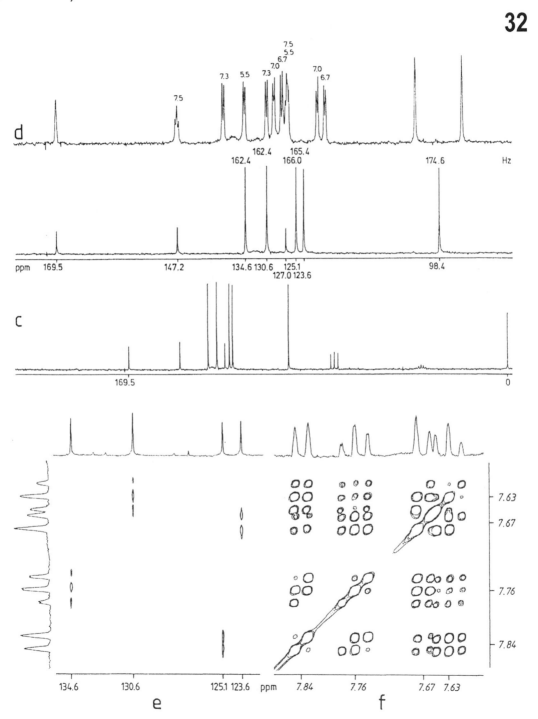

Problem 33

Commercial cyclopentadiene produces the set of NMR spectra **33**. In what form does this compound actually exist, and what is its relative configuration?

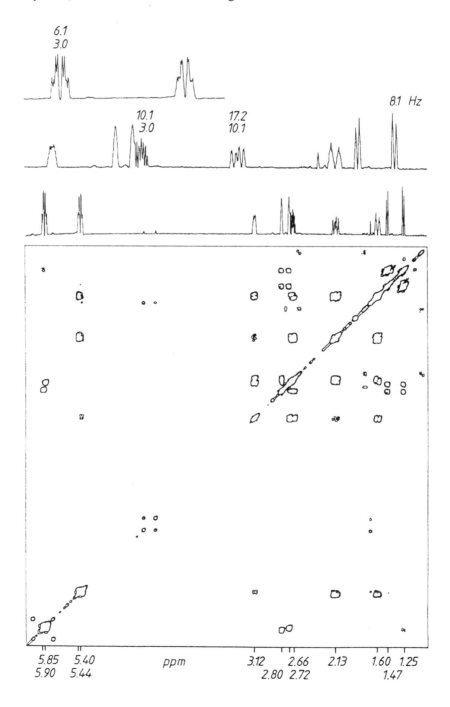

Problem 33, continued

Conditions: $(CD_3)_2CO$, 95% v/v, 25 °C, 400 and 200 MHz (^1H), 100 MHz (^{13}C). (**a**) ^1H NMR spectra (200 MHz) with expansion (above) and *HH* COSY diagram (below); (**b**) NOE difference spectra (200 MHz) with decoupling at $\delta_H = 1.25$ and *1.47*; (**c,d**) *CH* COSY contour plots with DEPT subspectra to distinguish *CH* (positive) and CH_2 (negative); (**c**) alkyl shift range; (**d**) alkenyl shift range.

33

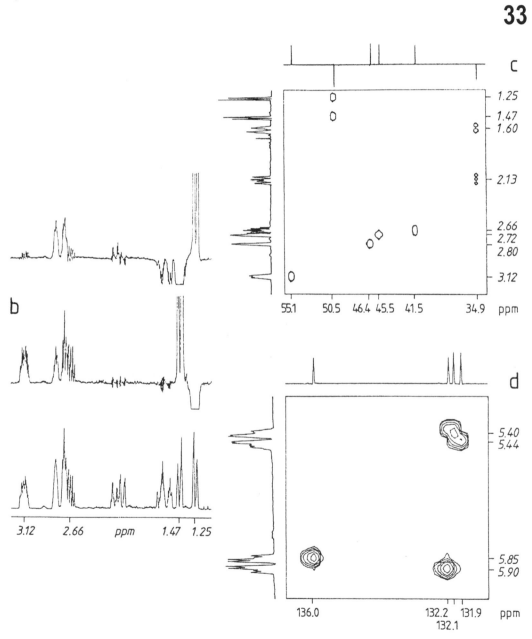

Problem 34

Several shifts and coupling constants in the NMR spectra set **34** are so typical that the carbon skeleton can be deduced without any additional information. An NOE difference spectrum gives the relative configuration of the compound.

34

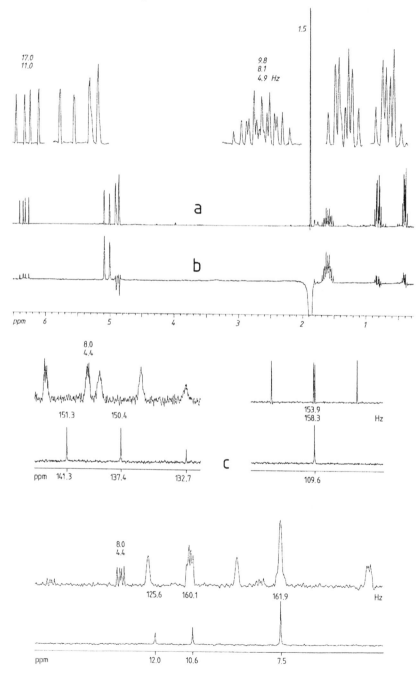

Problem 34, continued

Conditions: CDCl₃, 25°C, 200 MHz (¹H), 50 MHz (¹³C). (**a**) ¹H NMR spectrum with expanded multiplets; (**b**) NOE difference spectrum, irradiated at $\delta_H = 1.87$; (**c**) ¹³C NMR partial spectra, each with ¹H broadband decoupled spectrum below and NOE enhanced coupled spectrum (gated decoupling) above; (**d**) *CH* COSY diagram ('empty' shift ranges omitted).

34d

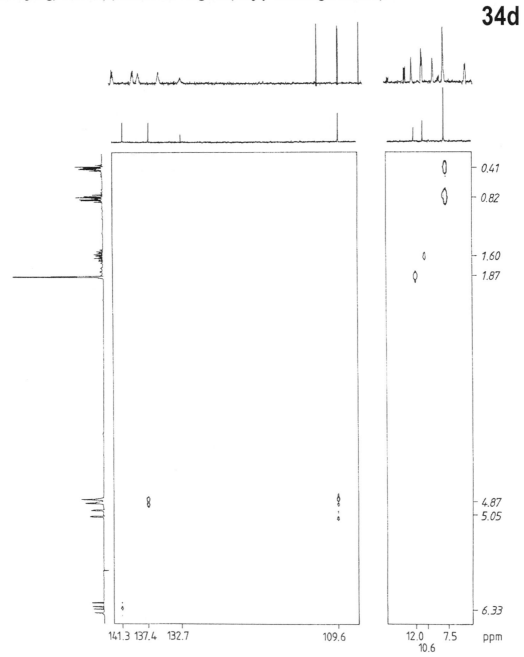

Problem 35

Which C*H* skeleton can be deduced from the NMR experiments **35**? What relative configuration does the ^1H NMR spectrum indicate?

Conditions: $(CD_3)_2CO$, 90% v/v, 25 °C, 400 MHz (^1H), 50 and 100 MHz (^{13}C). (**a**) Symmetrised INADEQUATE diagram (50 MHz); (**b**) C*H* COSY diagram with expansion (**c**) of the ^1H NMR spectrum between $\delta_H = 1.5$ and 2.3.

35a

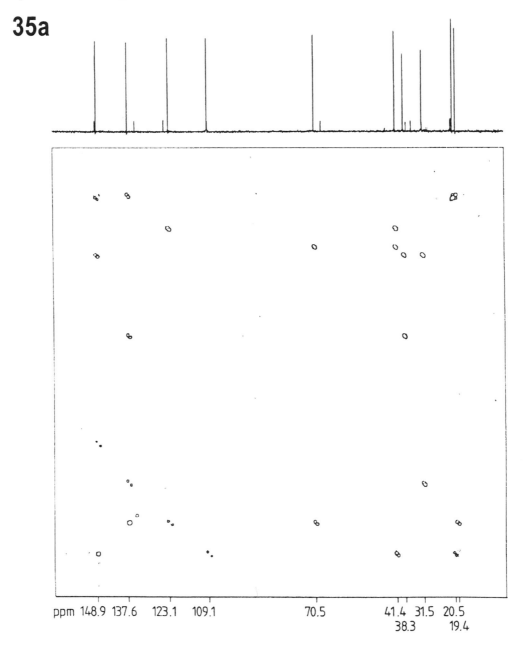

ppm 148.9 137.6   123.1   109.1   70.5   41.4 31.5 20.5
                                                                              38.3   19.4

Problem 35, continued

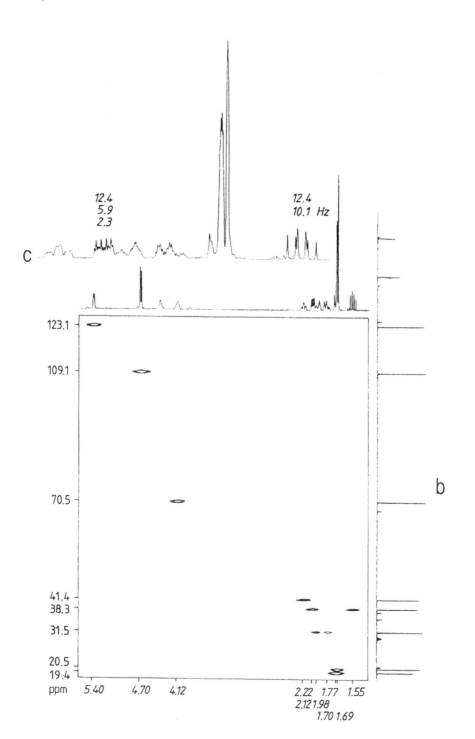

35

Problem 36

No further information is required to identify this compound and its relative configuration from the set of NMR experiments **36**?

Conditions: (CDCl$_3$, 50 % v/v, 25 °C, 400 MHz (^1H), 100 MHz (^{13}C). (**a**) Symmetrised INADE-QUATE diagram (100 MHz) with ^1H broadband decoupled ^{13}C NMR spectrum (**b**) and DEPT subspectra (**c**); (**d**) C*H* COSY diagram; (**e**) ^1H NMR spectra and NOE difference experiments with decoupling at $\delta_H = 0.74$ and 3.43.

36

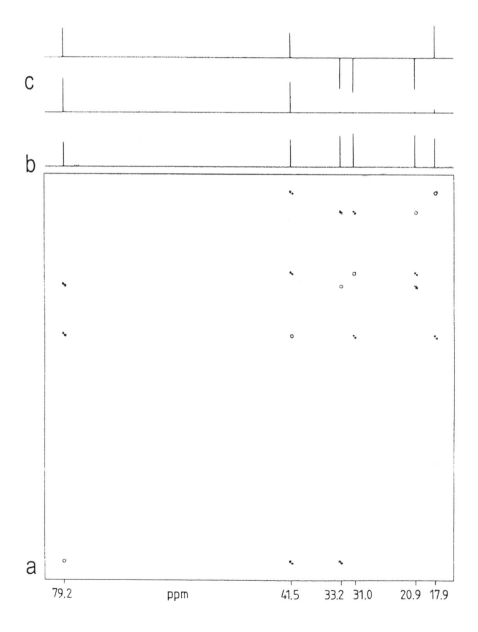

Problem 36, continued

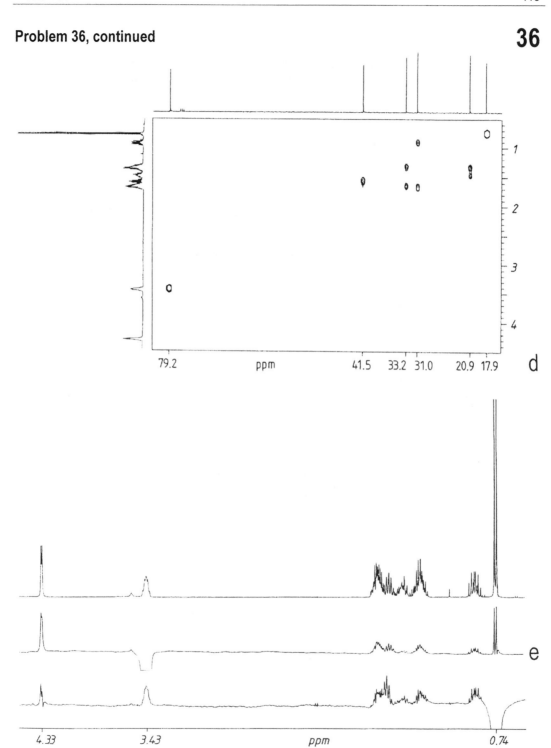

Problem 37

Cyclohexene oxide and metallated 2-methylpyridine reacted to afford a product which gave the NMR results **37**. Identify the relative configuration of the product and assign the resonances.

37

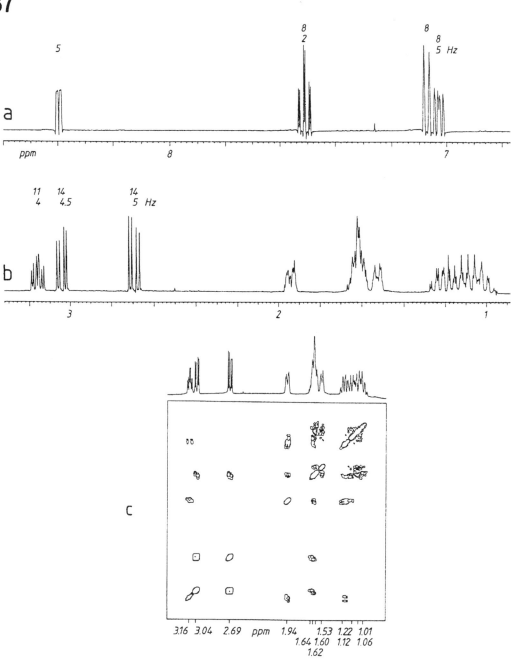

Problem 37, continued

Conditions: CDCl₃, 25°C, 400 MHz (¹H), 100 MHz (¹³C). (**a,b**) *¹H* NMR spectra, aromatic region
(**a**), aliphatic region (**b**); (**c**) *HH* COSY plot of aliphatic shift range; (**d**) C*H* COSY plot with
DEPT subspectra to distinguish C*H* and C*H₂*;

37d

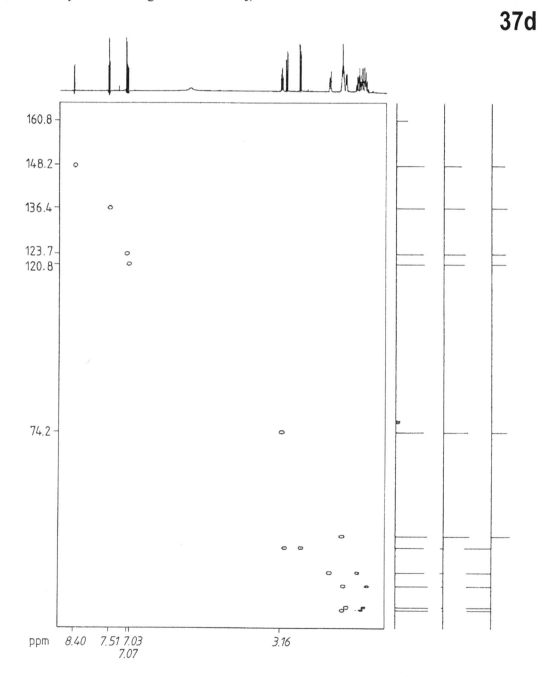

Problem 37, continued

(**e**) *CH* COSY diagram showing the region from $\delta_C = 24.7$ to 45.4; (**f**) symmetrised INADEQUA-TE plot of aliphatic region.

37

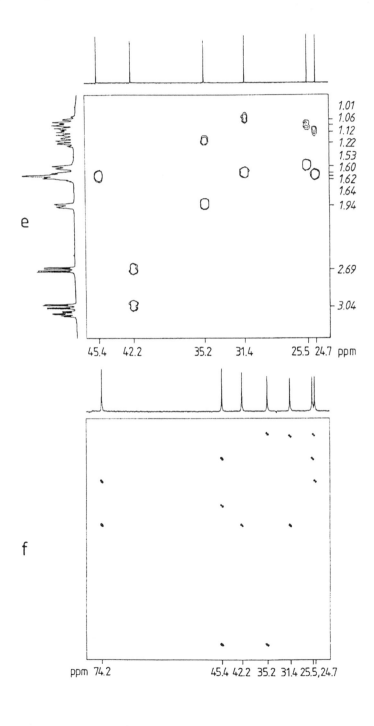

Problem 38

A fragrant substance found in cucumber and melon produces the NMR spectra set **38.** The identity and structure of the substance can be derived from these spectra without any further information.

Conditions: CDCl$_3$, 30 mg per 0.3 ml, 25 °C, 400 MHz (^{1}H), 100 MHz (^{13}C).

(**a**) *HH* COSY diagram;

38a

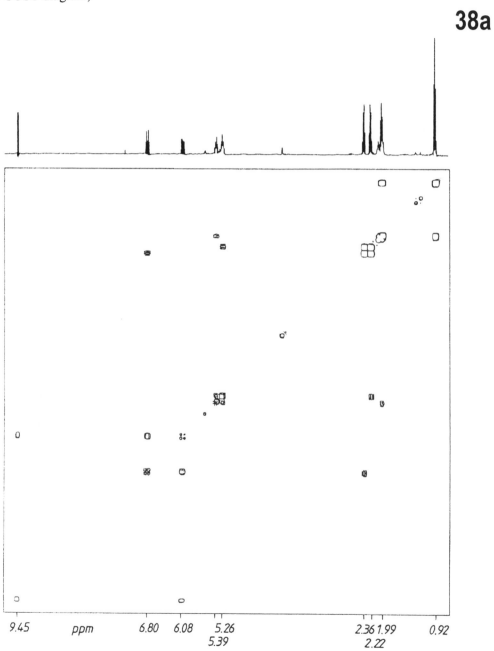

Problem 38, continued

(**b**) ^1H NMR spectrum with expanded multiplets; (**c**) ^{13}C NMR partial spectra, each with ^1H broadband decoupled spectrum below and NOE enhanced coupled spectrum (gated decoupling) above;

38

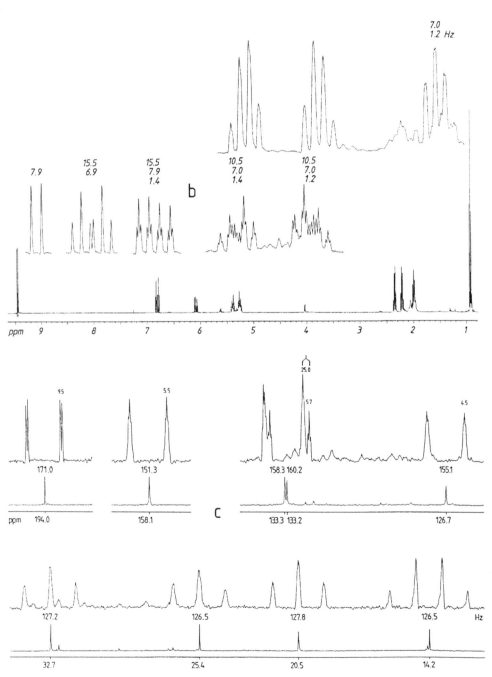

Problem 38, continued

(**d**) C*H* COSY diagram with expanded section ($\delta_C = 133.2 - 133.3$).

38d

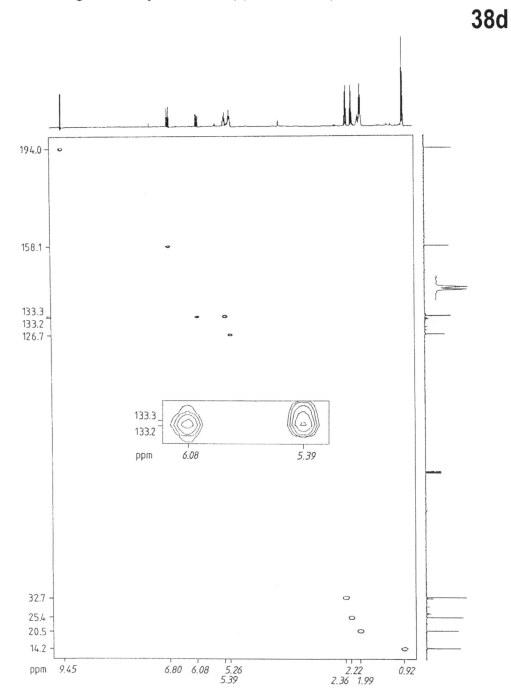

Problem 39

2,5-Bis(3,4-diethyl-2-pyrrolylmethyl)-3,4-diethyl-1*H*-pyrrole (*2*), prepared *in situ* from the di-*t*-butylester of the 5,5′-dicarboxylic acid (*1*), reacts with 4*H*-1,2,4-triazole-3,5-dialdehyde (*3*) in dichloromethane in the presence of trifluoroacetic acid and 2,3-dichloro-5,6-dicyano-*p*-benzoquinone as an oxidation reagent. Dark blue crystals are obtained after chromatographic purification. The dark violet chloroform solution fluoresces purple at 360 nm and gives the NMR experiments **39**. Which compound and which tautomer of it has been formed?

39a

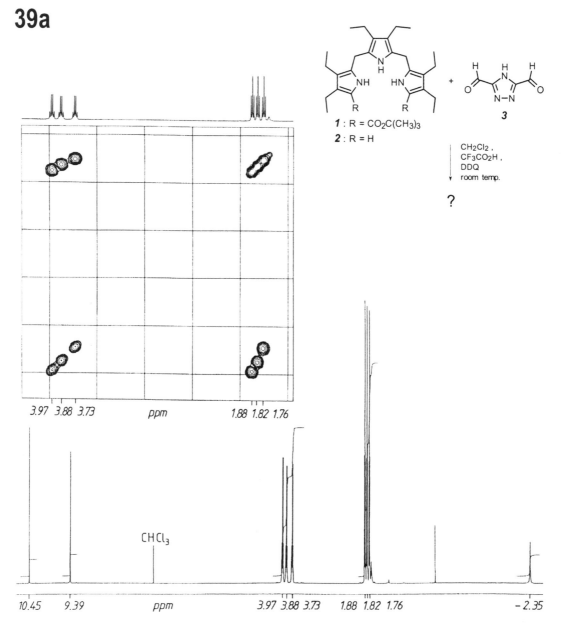

Problem 39, continued

Conditions: CDCl₃, 25 °C, 500 MHz (¹H), 125 MHz (¹³C). (**a**) ¹H NMR spectrum and *HH* COSY plot of ethyl groups; (**b**) *HC* HSQC plot with inserted zoomed section of ethyl groups;

39b

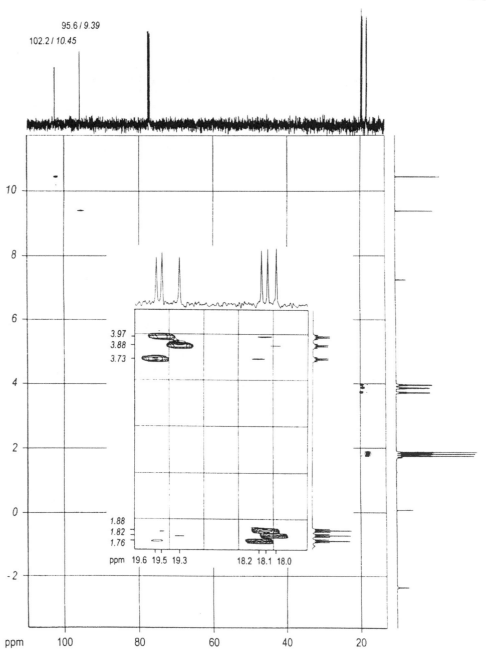

Problem 39, continued

(c) relevant sections of the *H*C HMBC plot.

39c

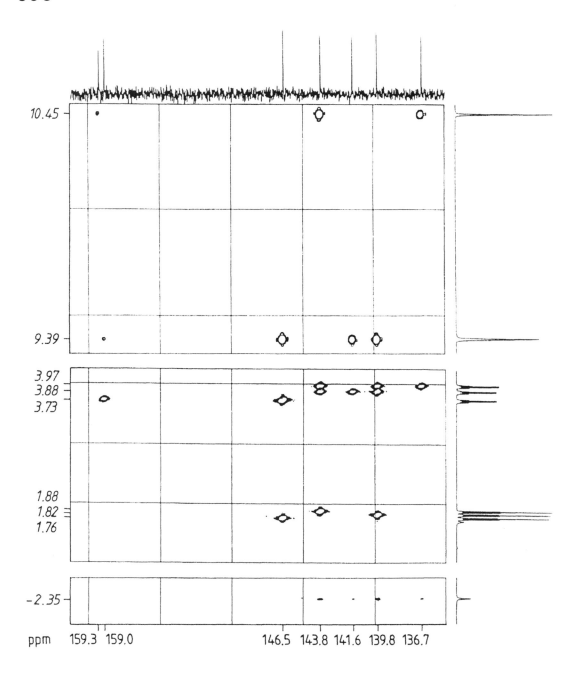

ppm   159.3 159.0   146.5 143.8 141.6 139.8 136.7

Problem 40

What compound $C_{18}H_{20}O_6$ can be identified from the CH correlation experiments **40** and the ^1H NMR spectra shown above?

Conditions: $(CD_3)_2SO$, 25 °C, 200 MHz (^1H), 50 MHz (^{13}C). CH COSY (shaded contours) and CH COLOC plot (unshaded contours) in one diagram; in the ^1H NMR spectrum the signal at $\delta_H = 12.34$ disappears following D_2O exchange.

40

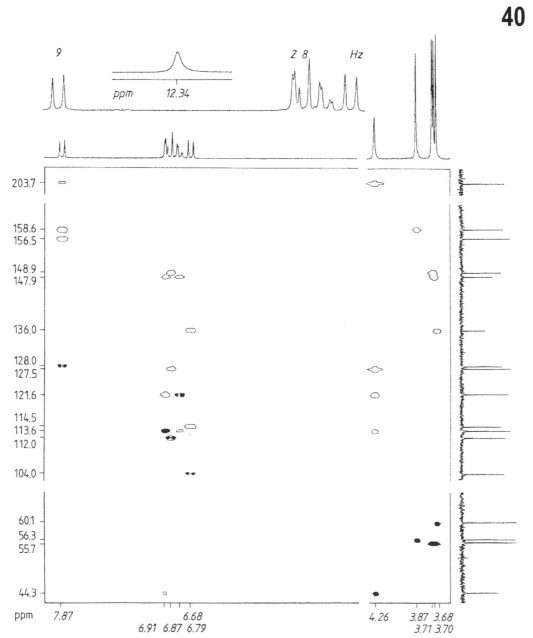

Problem 41

What compound $C_{19}H_{18}O_6$ can be identified from the CH COSY and CH COLOC plots **a** with the ^1H NMR spectra shown above and from the ^{13}C NMR spectra **b**?

41a

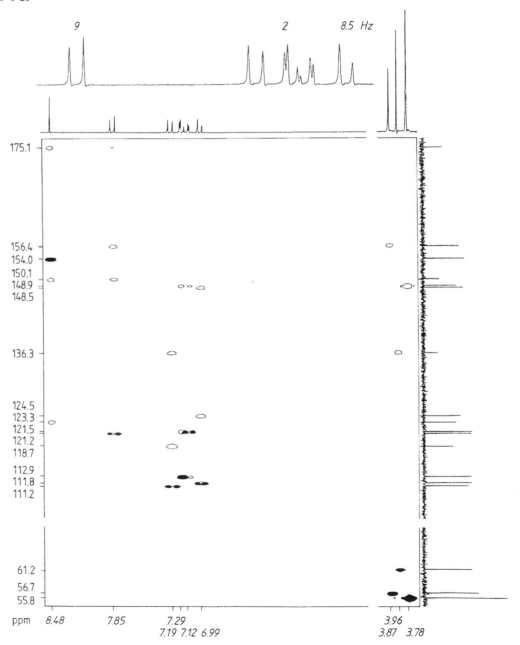

Problem 41, continued

Conditions: $(CD_3)_2SO$, 25 °C, 200 MHz (^1H), 50 MHz (^{13}C).

(**a**) C*H* COSY diagram (shaded contours) and C*H* COLOC spectra (unshaded contours) in one diagram with an expansion of the ^1H NMR spectrum; (**b**) parts of ^{13}C NMR spectra to be assigned, with ^1H broadband decoupled spectrum below and NOE enhanced coupled spectrum (gated decoupling) above.

41b

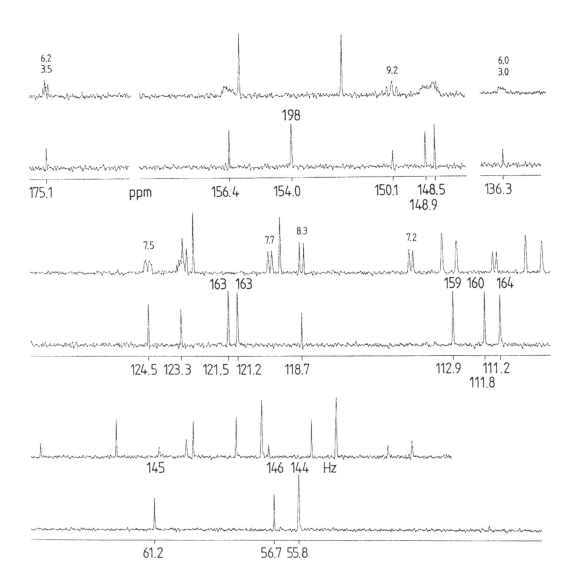

Problem 42

3',4',6,7-Tetramethoxyisoflavone (*3*) was the target of the cyclisation reaction of 3,4-dimethoxy-phenol (*1*) with formyl-(3,4-dimethoxyphenyl)acetic acid (*2*) in the presence of polyphosphoric acid.

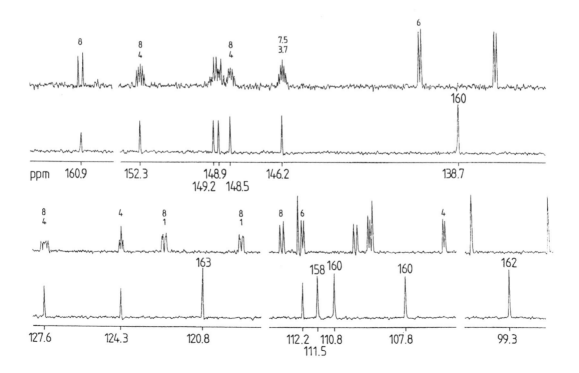

A pale yellow, crystalline product is obtained which fluoresces intense blue and gives the NMR results **42**. Does the product have the desired structure?

Conditions: CDCl$_3$, 25 °C, 200 MHz (^1H), 50 MHz (^{13}C). (**a**) C*H* COSY (shaded contours) and C*H* COLOC diagrams (unshaded contours) in one diagram with enlarged section (**b**), and with expanded methoxy quartets (**c**); (**d**) sections of ^{13}C NMR spectra, each with ^1H broadband decoupled spectrum below and NOE enhanced coupled spectrum (gated decoupling) above.

42d

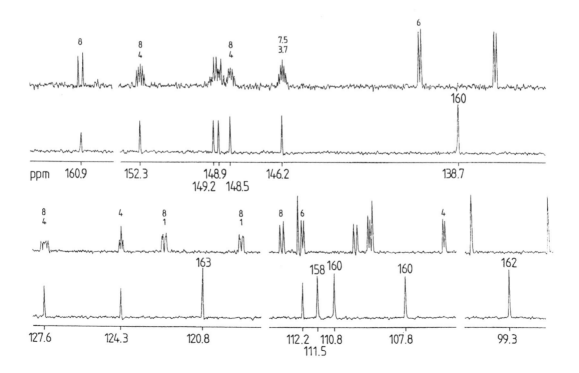

Problem 42, continued

42

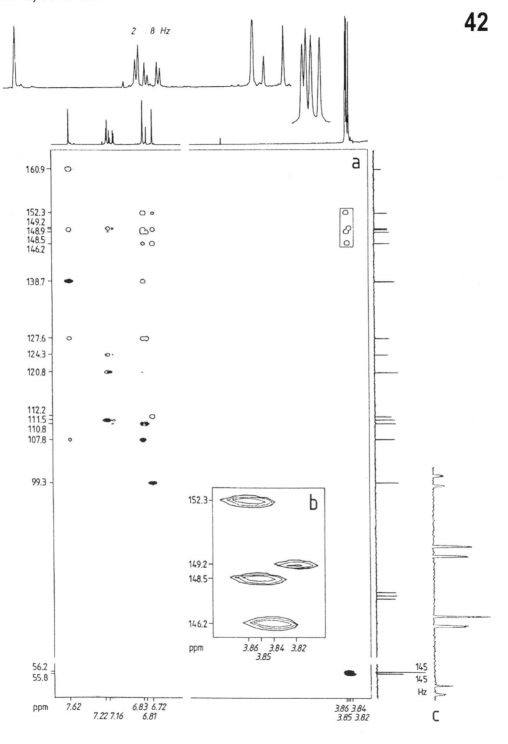

Problem 43

An aflatoxin is isolated from *Aspergillus flavus*. Which of the three aflatoxins, B_1, G_1 or M_1, is it given the set of NMR experiments **43**?

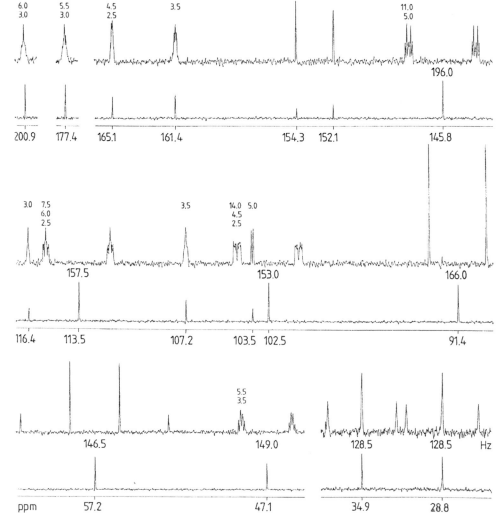

Aflatoxin B_1

Aflatoxin G_1

Aflatoxin M_1

Conditions: $(CD_3)_2SO$, 25 °C, 200 MHz (^1H), 50 MHz (^{13}C). (**a**) Sections of ^{13}C NMR spectra, in each case with ^1H broadband decoupled spectrum below and NOE enhanced spectrum (gated decoupling) above;

43a

Problem 43, continued

(**b**) *CH* COSY (shaded contours) and *CH* COLOC plots (unshaded contours) in one diagram with expanded ^1H multiplets.

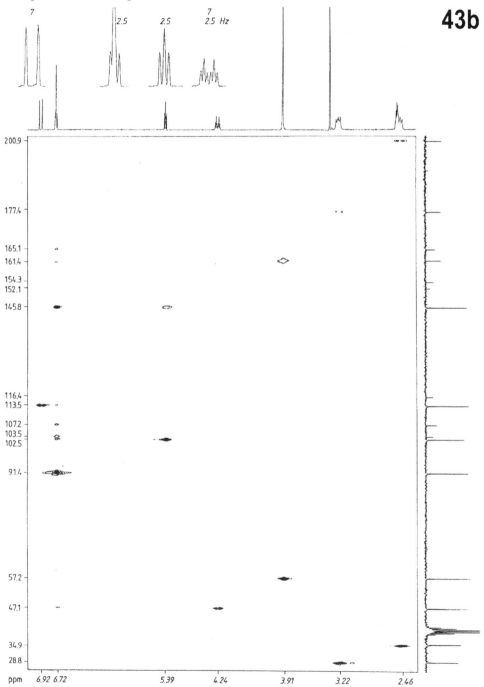

Problem 44

The degradation of 2,6-xylenol (2,6-dimethylphenol) by bacteria produces a metabolite with elemental composition $C_8H_{10}O_2$ as determined by high-resolution mass spectrometry [39]. Which carbon skeleton and which relative configuration are deducible from the NMR experiments **44**, all obtained from one 1.5 mg sample?

Conditions: $CDCl_3$, 25 °C, 400 MHz (^1H), 100 MHz (^{13}C). (**a**) *HH* COSY plot with ^1H NMR spectrum and *HH* coupling constants; the signal at $\delta_H = 2.17$ belongs to acetone (impurity).

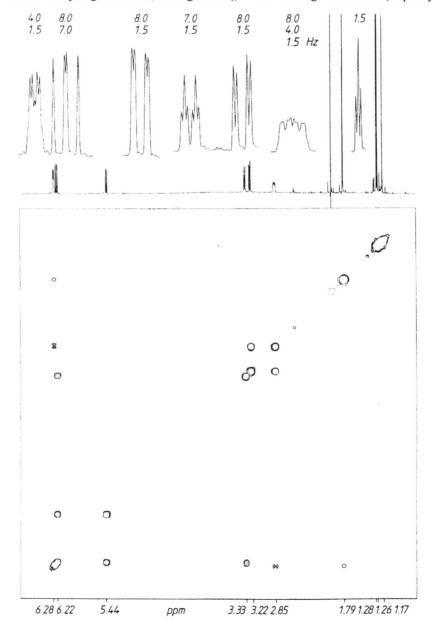

Problem 44, continued

(**b**) *CH* COSY plot with DEPT subspectra for analysis of the *CH* multiplicities;

44b

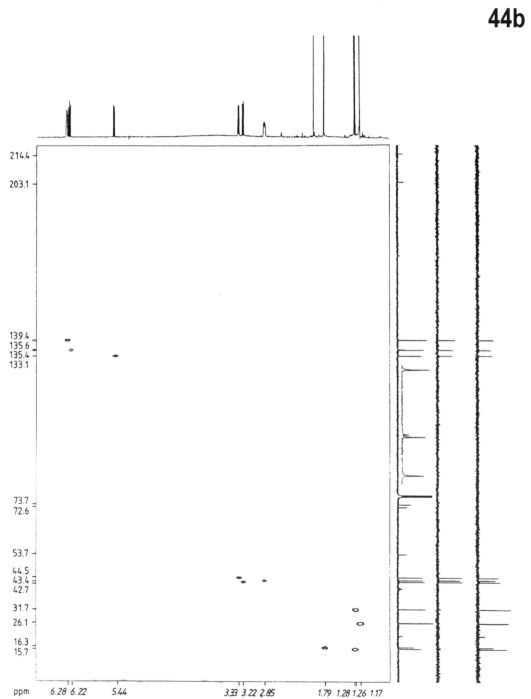

Problem 44, continued

(c) *CH* COLOC plot, section of methyl protons;

44c

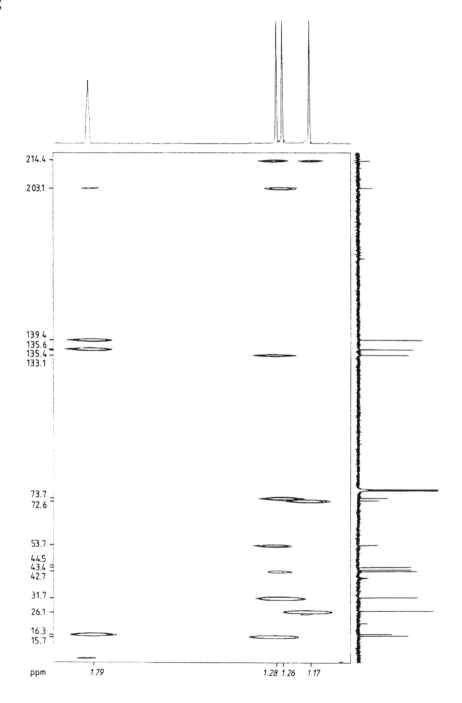

Problem 44, continued

(**d**) NOE difference spectra, irradiation of all protons.

44d

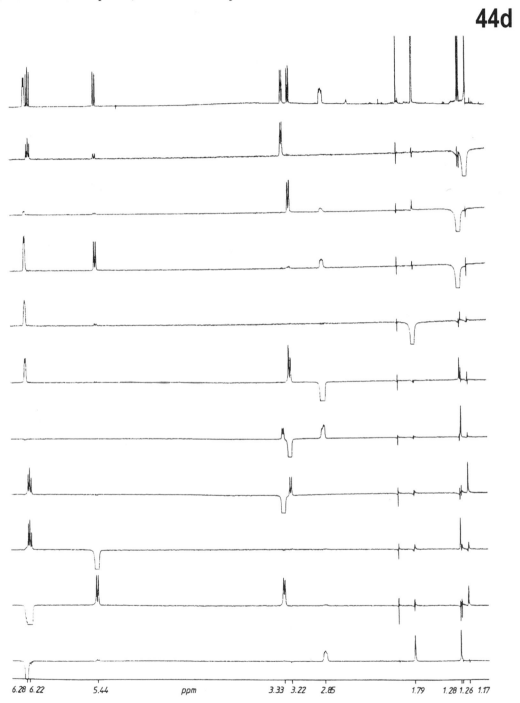

6.28 6.22   5.44   ppm   3.33 3.22   2.85   1.79   1.28 1.26 1.17

Problem 45

From the plant *Escallonia pulverulenta* (Escalloniaceae), which grows in Chile, an iridoid gluco-side of elemental composition $C_{18}H_{22}O_{11}$ was isolated. Formula *1* gives the structure of the iridoid glucoside skeleton [40].

30 mg of the substance was available and was used to obtain the set of NMR results **45**. What structure does this iridoside have?

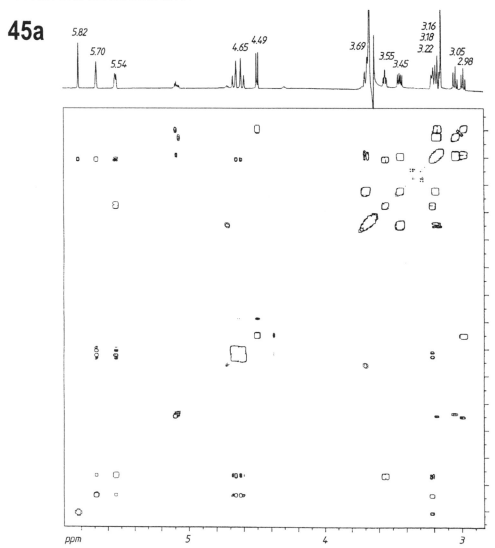

Problem 45, continued

Conditions: (CD$_3$)$_2$SO, 25 °C, 400 and 600 MHz (^1H), 100 MHz (^{13}C). (**a**) *HH* COSY plot (600 MHz) following D$_2$O exchange; (**b**) ^1H NMR spectra before and after deuterium exchange; (**c**) sections of the ^{13}C NMR spectra, in each case with the ^1H broadband decoupled spectrum below and NOE enhanced coupled spectrum (gated decoupled) above;

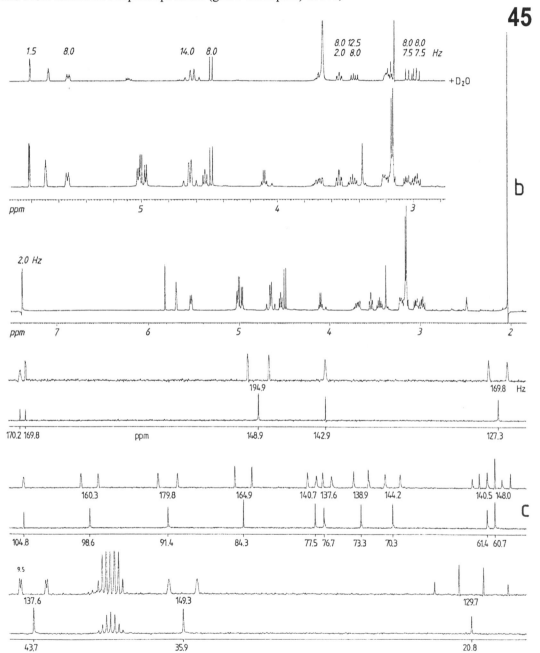

45

Problem 45, continued

(**d**) *CH* COSY plot with DEPT subspectra for analysis of the *CH* multiplicities;

45d

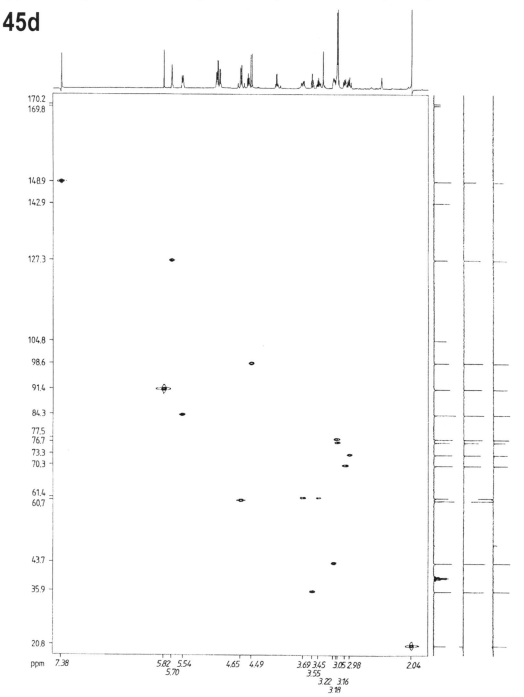

Problem 45, continued

(**e**) *CH* COLOC diagram.

45e

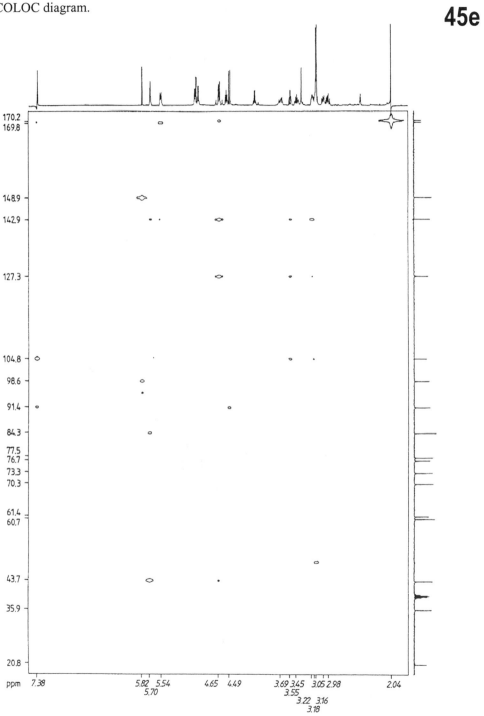

Problem 46

Hydrolysis of an oligosaccharide isolated from human milk affords D-galactose, D-glucose and D-glucosamine. For NMR analysis, the oligosaccharide is reduced so that the terminal sugar exists as polyol and does not mutarotate anymore. The reduction product is peracetylated so that the NMR experiments **46** can be recorded in deuteriochloroform solution. What is the sequence and the relative configuration of the parent oligosaccharide?

Conditions: CDCl$_3$, 25 °C, 500 MHz (^1H), 125 MHz (^{13}C); acetyl carbonyl (δ_C = 169-171.4) and acetyl methyl resonances (δ_C = 20.6 - 23.8, δ_H = 2.1 - 2.2) are not shown. (**a**) *HH* COSY plot with ^1H NMR spectrum;

46a

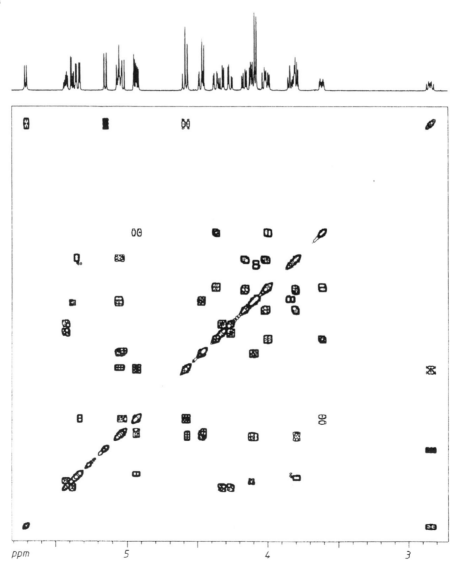

ppm   5   4   3

Problem 46, continued

(**b**) *H*C HSQC plot with ^1H NMR spectrum;

46b

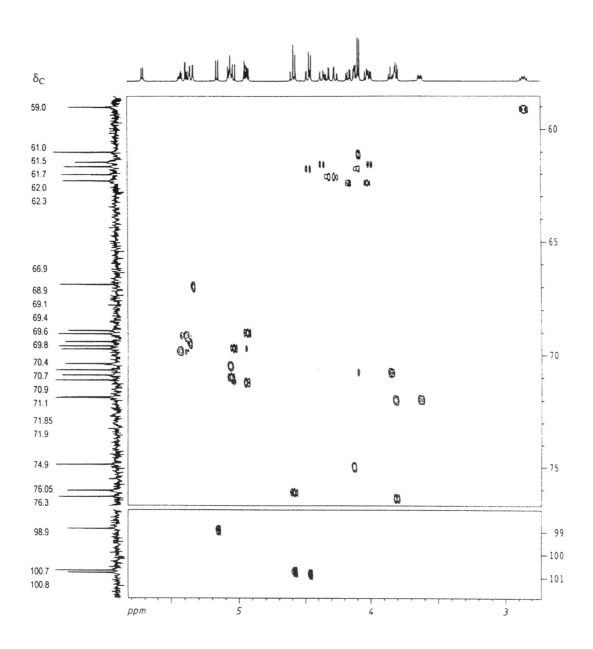

Problem 46, continued

46c

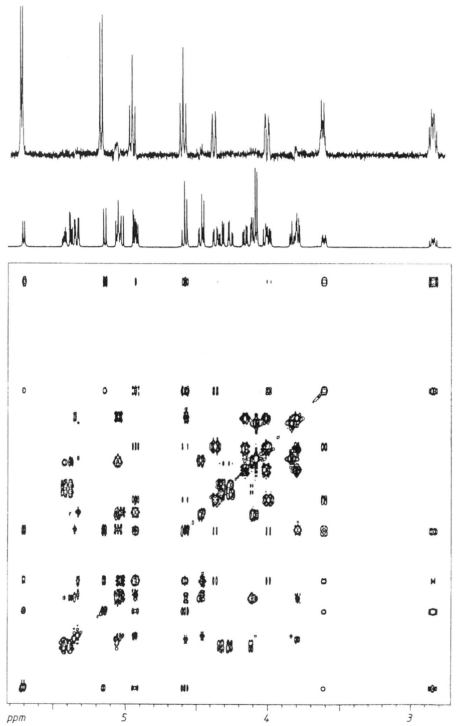

Problem 46, continued

(**c**) *HH* TOCSY plot with ^1H NMR spectrum and selective *HH* TOCSY with soft pulse irradiation at $\delta_H = 5.70$ (top); (**d**) *HH* ROESY plot with ^1H NMR spectrum, same section and same scales as shown in (**c**).

46d

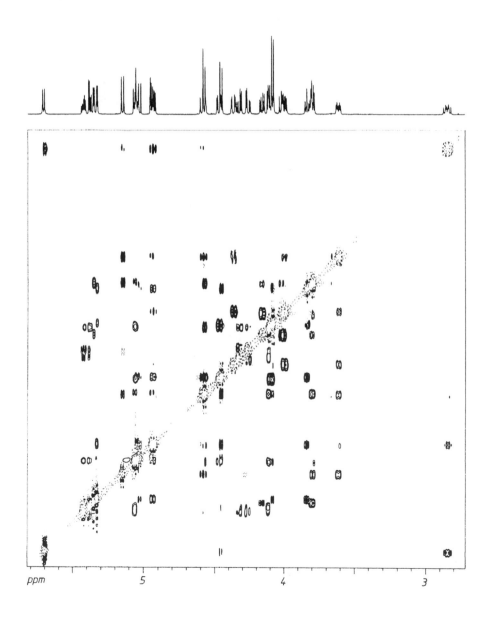

Problem 47

A compound with the elemental composition $C_{15}H_{22}O_3$ was isolated from the methanol extract of the Chilean medicinal plant *Centaurea chilensis* (Compositae) [41]. What is the structure of the natural product, given the NMR experiments **47**?

Conditions: CD$_3$OD, 15 mg per 0.3 ml, 400 MHz (^1H), 100 MHz (^{13}C). (**a**) *HH* COSY plot from $\delta_H = 1.2$ to *3.5*;

47a

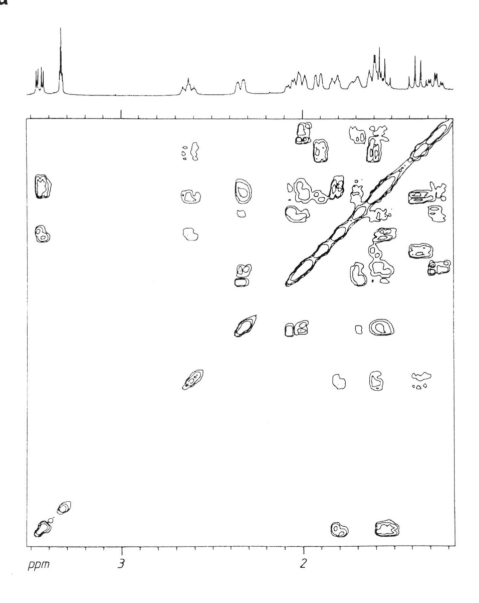

ppm   3   2

Problem 47, continued

(**b**) expanded ^1H multiplets from $\delta_H = 1.23$ to 3.42;

47b

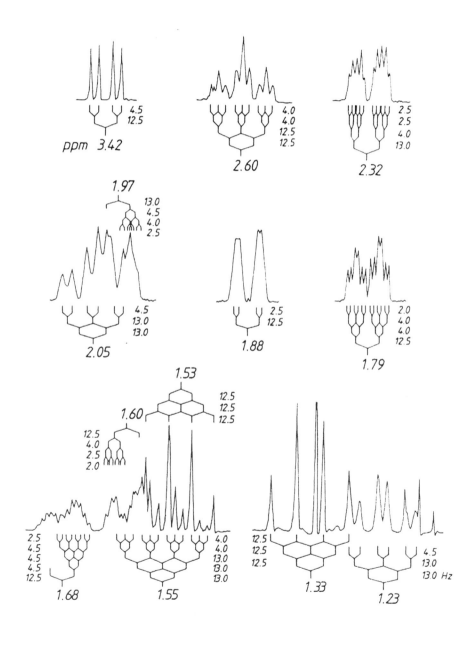

Problem 47, continued

(c) *CH* COSY diagram from δ_C = 6 to 130 with ^1H broadband decoupled ^{13}C NMR spectrum (**d**), DEPT *CH* subspectrum (**e**) and DEPT subspectrum (**f**, *CH* and *CH₃* positive, *CH₂* negative);

47

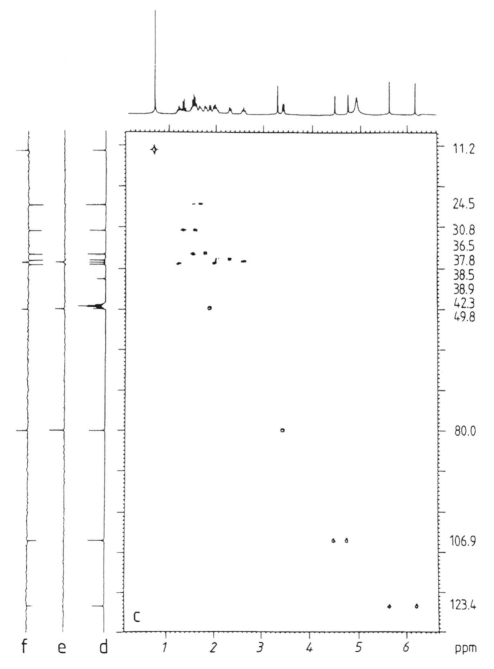

Problem 47, continued

(g) C*H* COLOC plot.

47g

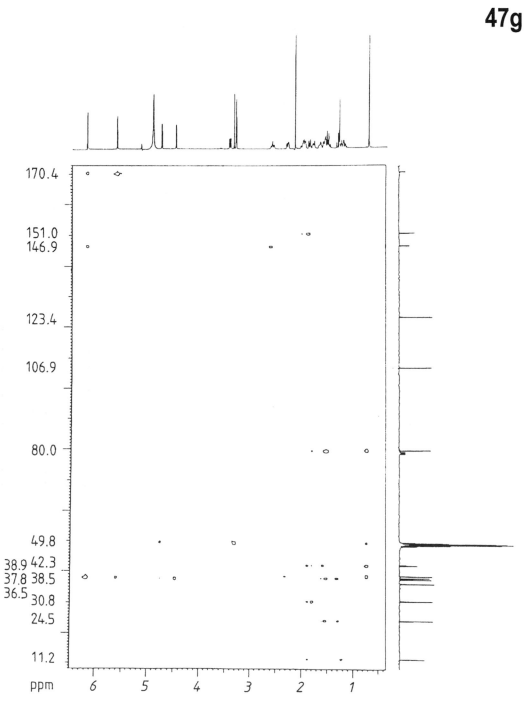

Problem 48

The umbelliferone ether structure *1* was suggested for a natural product which was isolated from galbanum resin [42]. Does this structure fit the NMR results **48**? Is it possible to give a complete spectral assignment despite lack of resolution of the proton signals at 200 MHz? What statements can be made about the relative configuration?

48a

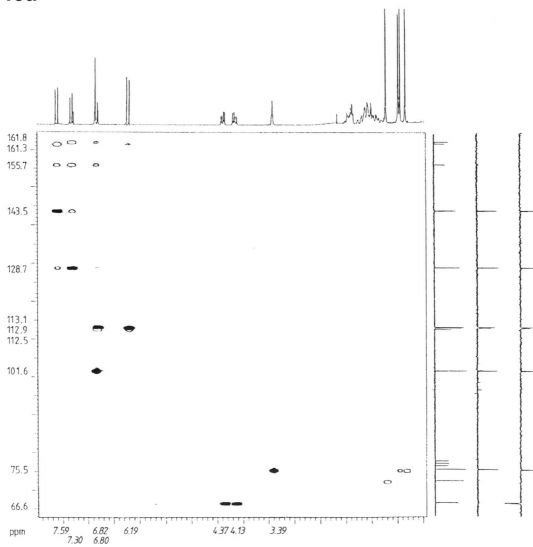

Problem 48, continued

Conditions: CDCl$_3$, 50 mg per 0.3 ml, 25 °C, 200 and 400 MHz (^1H), 50 and 100 MHz (^{13}C). (**a,b**) C*H* COSY(shaded contours) and C*H* COLOC plots (unshaded contours) in one diagram with DEPT subspectra for identification of the C*H* multiplets; (**a**) ^{13}C shift range from δ_C = 66.6 to 161.8; (**b**) ^{13}C shift range from δ_C = 16.0 to 75.5;

48b

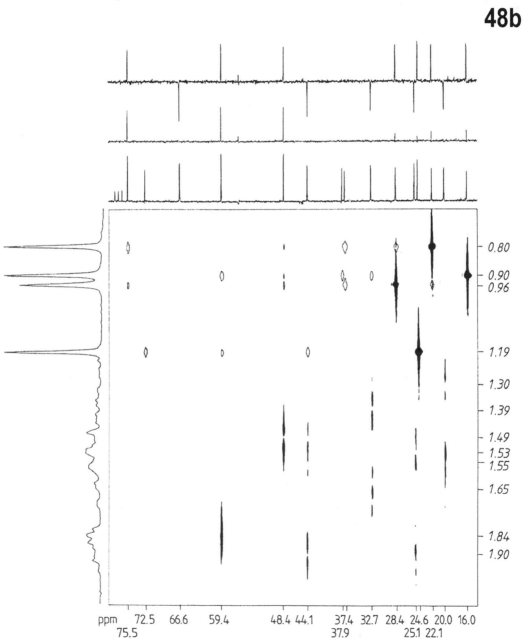

Problem 48, continued

(c) sections of ^{13}C NMR spectra (100 MHz), ^1H broadband decoupled spectrum below and NOE enhanced coupled spectrum (gated decoupling) above, with expanded multiplets in the aromatic range;

48c

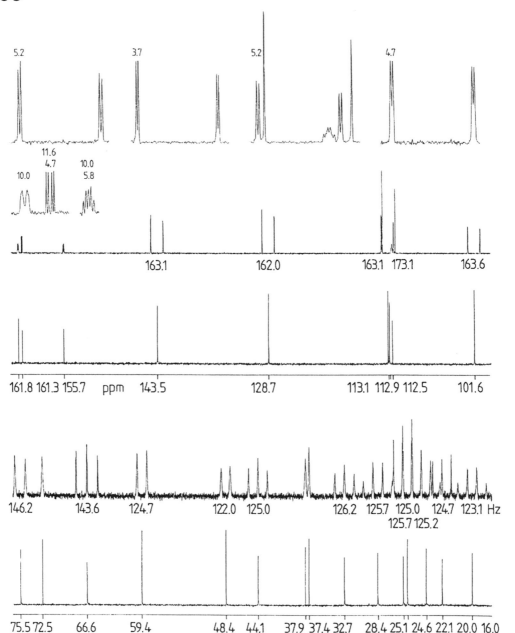

Problem 48, continued

(**d**) ^1H NMR spectrum with expanded multiplets, integral and NOE difference spectra (200 MHz, irradiated at δ_H = 0.80, 0.90, 0.96, 1.19, 3.39 and 4.13 / 4.37).

48d

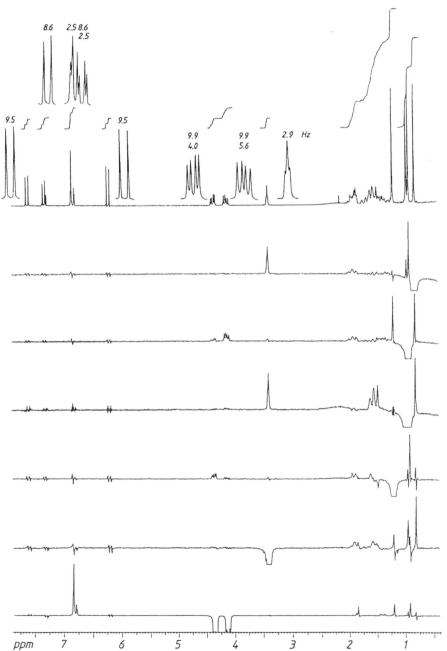

Problem 49

A natural product isolated from the plant *Euryops arabicus*, native to Saudi Arabia, has the elemental composition $C_{15}H_{16}O$ determined by mass spectrometry. What is its structure, given the NMR experiments **49**?

Conditions: $CDCl_3$, 20 mg per 0.3 ml, 25 °C, 400 MHz (^1H), 100 MHz (^{13}C).

(**a**) ^1H NMR spectrum with expanded multiplets;

49a

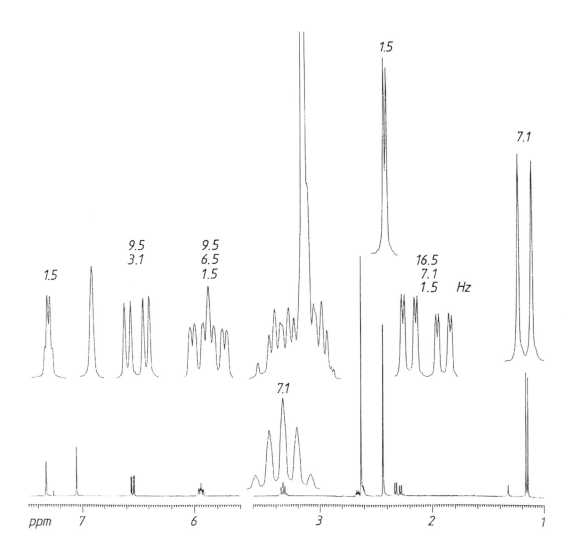

Problem 49, continued

(**b**) sections of ^{13}C NMR spectra, in each case with ^1H broadband decoupled spectrum below and NOE enhanced coupled spectrum (gated decoupling) above;

49b

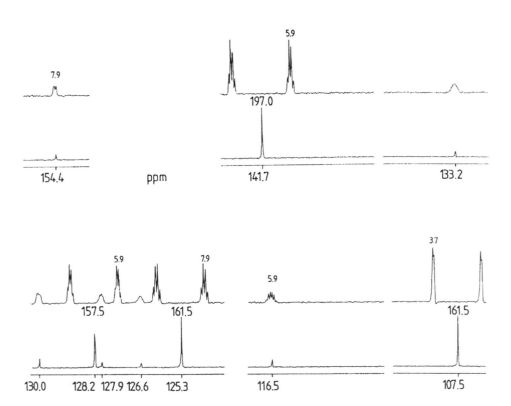

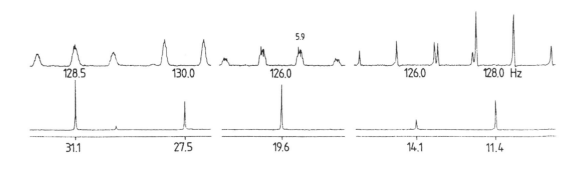

Problem 49, continued

(c) C*H* COSY and C*H* COLOC plots in one diagram with DEPT subspectra to facilitate analysis of the C*H* multiplicities;

49c

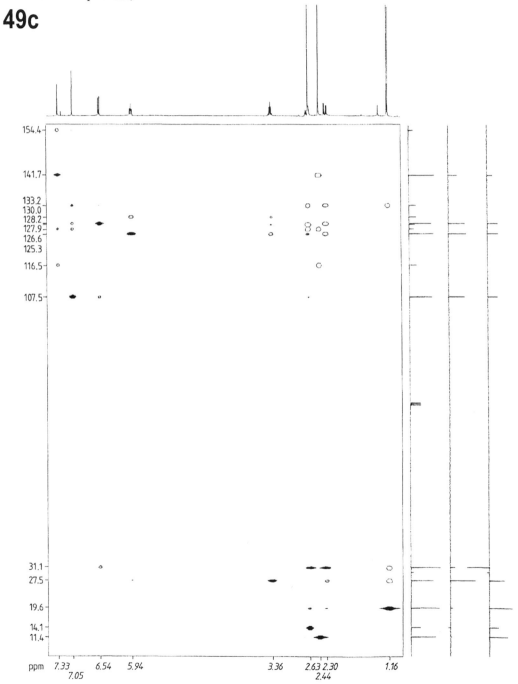

Problem 49, continued

(**d**) enlarged section of (**c**).

49d

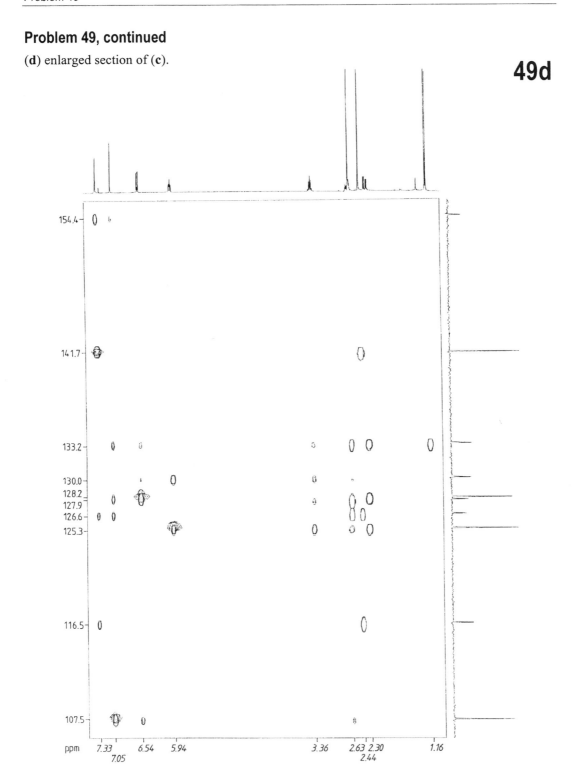

Problem 50

A compound with the elemental composition $C_{21}H_{28}O_6$ determined by mass spectrometry, was isolated from the light petroleum extract of the leaves of *Senecio darwinii* (Compositae, Hooker and Arnolt) [43], a plant which grows in Tierra del Fuego (Chile). What structure can be derived from the set of NMR experiments **50**?

Conditions: $CDCl_3$, 25 mg per 0.3 ml, 25°C, 400 MHz (^1H), 100 MHz (^{13}C).

(**a**) *HH* COSY plot;

50a

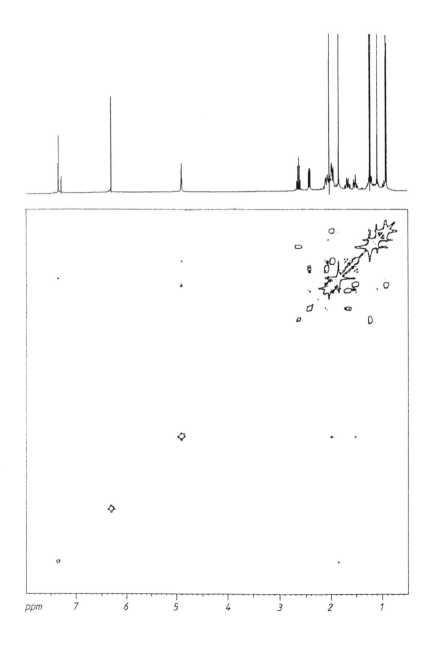

ppm   7   6   5   4   3   2   1

Problem 50, continued

(**b**) *HH* COSY plot, section from $\delta_H = 0.92$ to *2.62*;

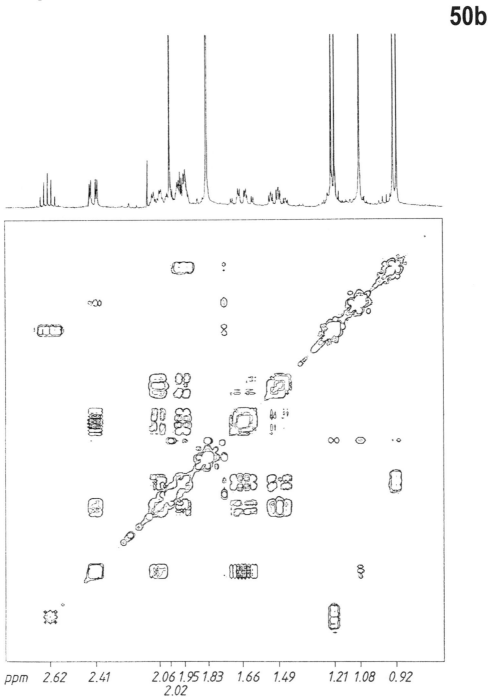

ppm   *2.62*   *2.41*   *2.06* *1.95* *1.83*   *1.66*   *1.49*   *1.21* *1.08*   *0.92*
2.02

Problem 50, continued

(**c**) C*H* COSY plot with DEPT subspectra to facilitate analysis of the C*H* multiplets;

50c

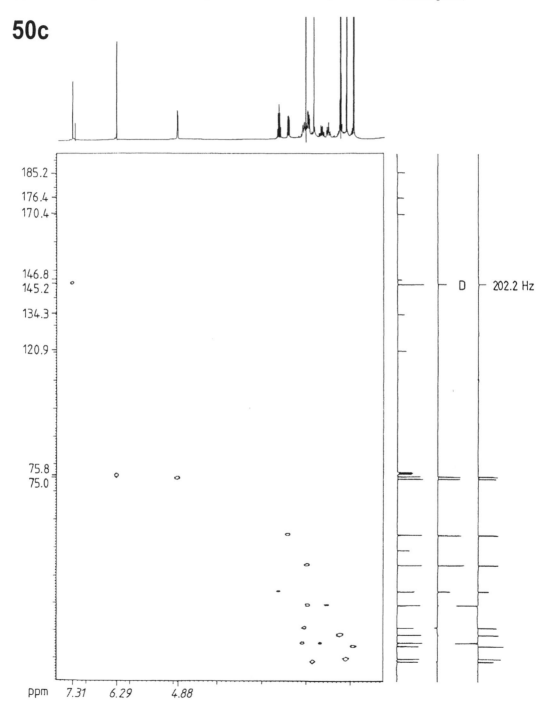

Problem 50, continued

(**d**) *CH* COSY plot, sections from $\delta_H = 0.92$ to 2.62 and $\delta_C = 8.8$ to 54.9;

50d

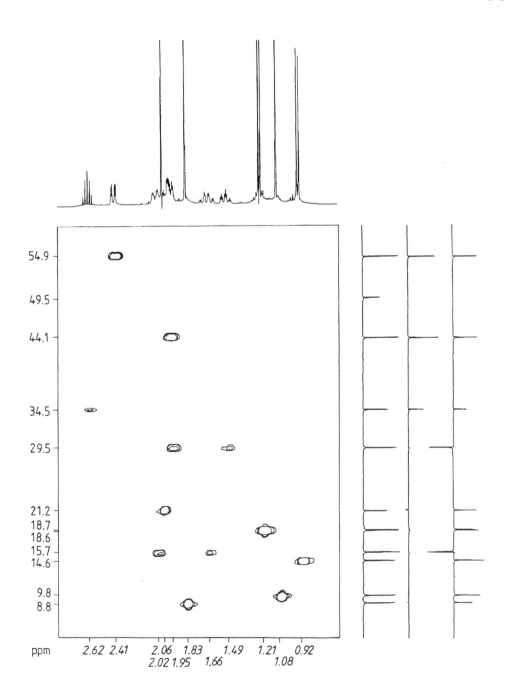

ppm   2.62 2.41   2.06 1.83   1.49   1.21   0.92
             2.02 1.95   1.66   1.08

Problem 50, continued

(e) C*H* COLOC plot;

50e

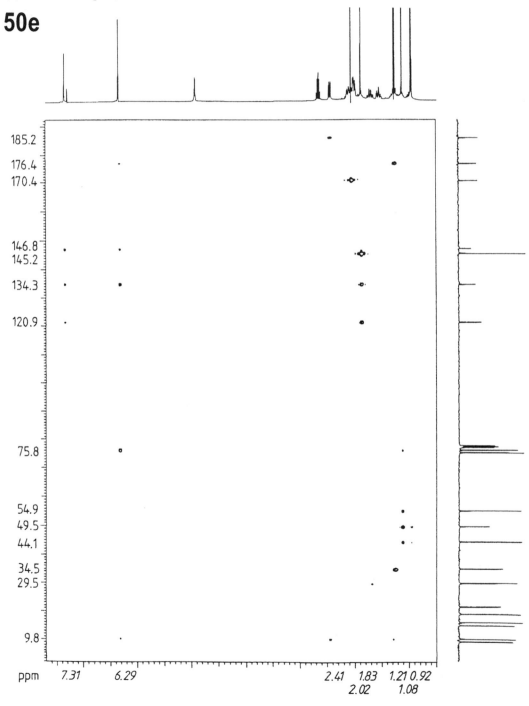

Problem 50, continued

(**f**) ^1H NMR spectrum with expanded multiplets and NOE difference spectra, irradiations at $\delta_H =$ *1.49, 1.66, 2.41* and *6.29*.

50f

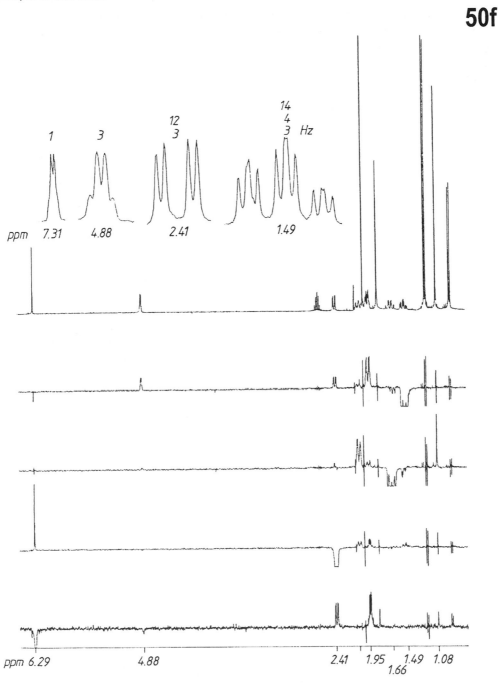

Problem 51

Sapogenins of the dammaran type were isolated from the leaves and roots of the plant *Panax notoginseng*, native to China [44]. One of these sapogenins has the elemental composition $C_{30}H_{52}O_4$ and produces the set of NMR results **51**. What is the structure of the sapogenin?

Conditions: 20 mg, $CDCl_3$, 20 mg per 0.3 ml, 25 °C, 200 and 400 MHz (^1H), 100 MHz (^{13}C).

(**a**) *HH* COSY plot (400 MHz) with expansion of multiplets;

51a

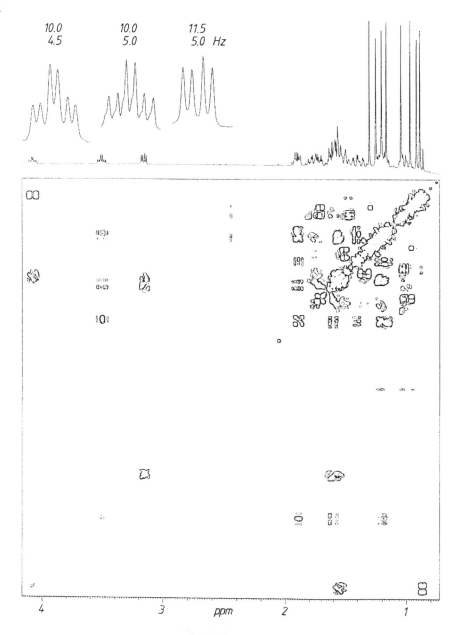

Problem 51, continued

(**b**) ^1H NMR and NOE difference spectra (200 MHz), decoupling of the methyl protons shown;

51b

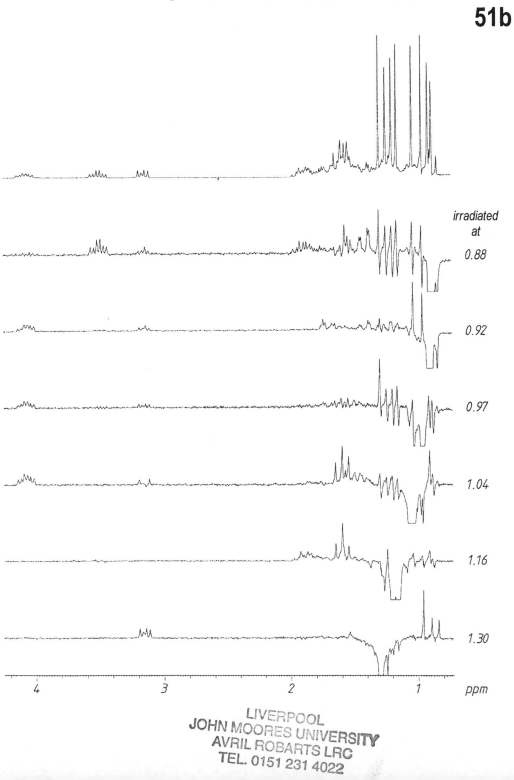

irradiated
at

0.88

0.92

0.97

1.04

1.16

1.30

ppm

Problem 51, continued

(c) C*H* COSY plot with enlarged section (**d**);

51c

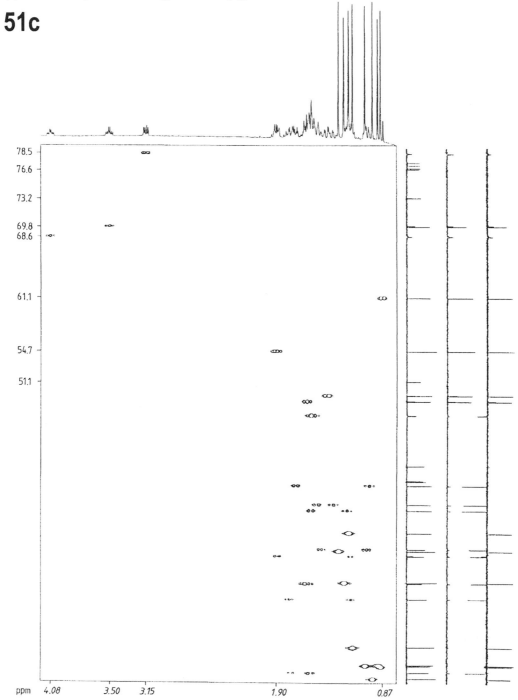

Problem 51, continued

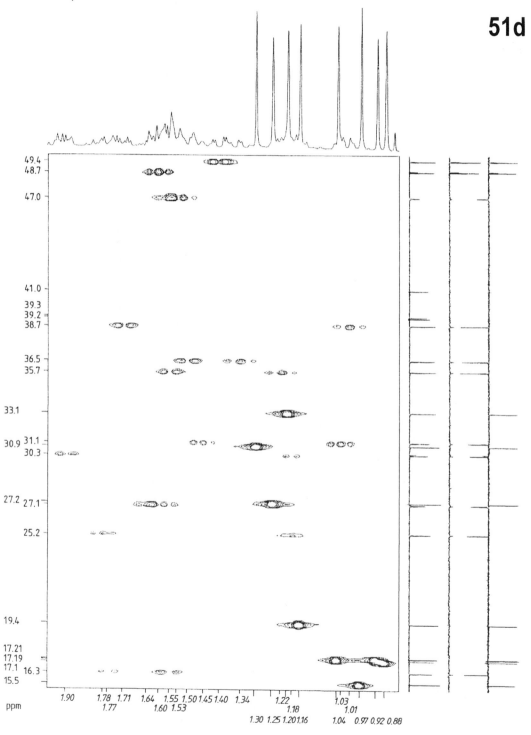

51d

Problem 51, continued

(**e**) relevant section of C*H* COLOC plot.

51e

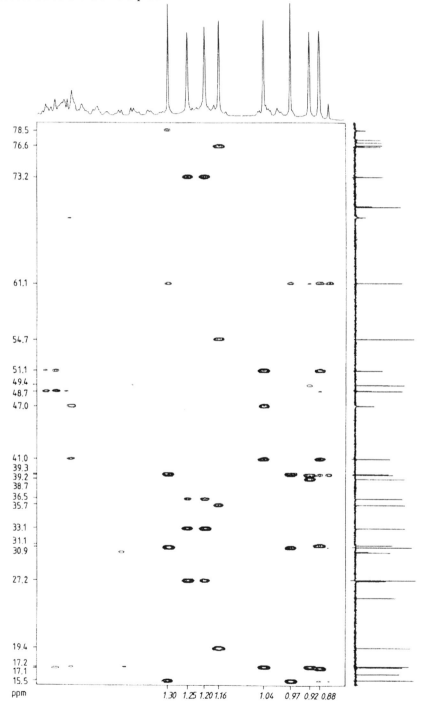

Problem 52

A compound with the molecular formula $C_7H_{13}NO_3$, determined by mass spectrometry, was isolated from the plant *Petiveria alliacea* (Phytolaccaceae). What is its structure given the set of NMR results **52**?

Conditions: CD$_3$OD, 30 mg per 0.3 ml, 25 °C, 400 and 200 MHz (^1H), 100 MHz (^{13}C).

(a) *HH* COSY plot (400 MHz) with expanded ^1H multiplets;

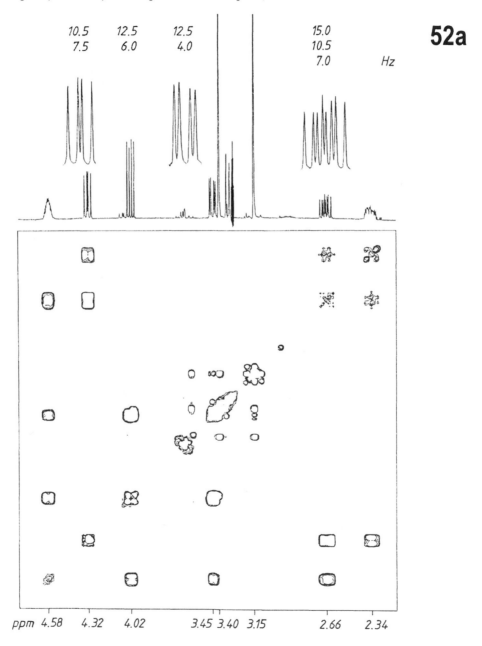

Problem 52, continued

(**b**) *CH* COSY and *CH* COLOC plots in one diagram with DEPT spectrum (*CH₂* negative, *CH* and *CH₃* positive) and coupled (gated decoupling) ^{13}C NMR spectrum above;

52b

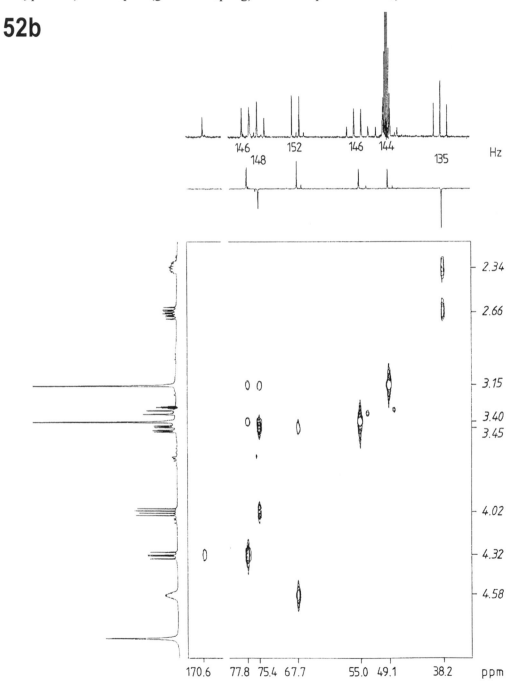

Problem 52, continued

(**c**) ^1H NOE difference spectra, irradiation at the signals indicated, 200 MHz.

52c

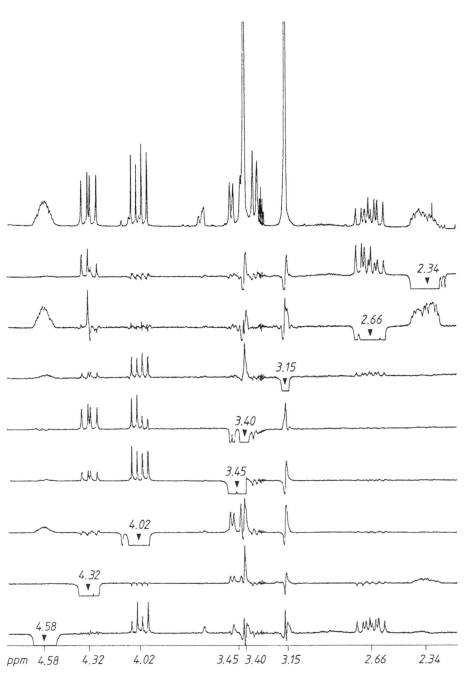

ppm   4.58   4.32   4.02   3.45 3.40   3.15   2.66   2.34

Problem 53

The hydrochloride of a natural product which is intoxicating and addictive produced the set of NMR results **53**. What is the structure of the material? What additional information can be derived from the NOE difference spectrum?

Conditions: CD₃OD, 30 mg per 0.3 ml, 25°C, 400 MHz (¹H), 100 MHz (¹³C). (**a**) *HH* COSY plot;

53a

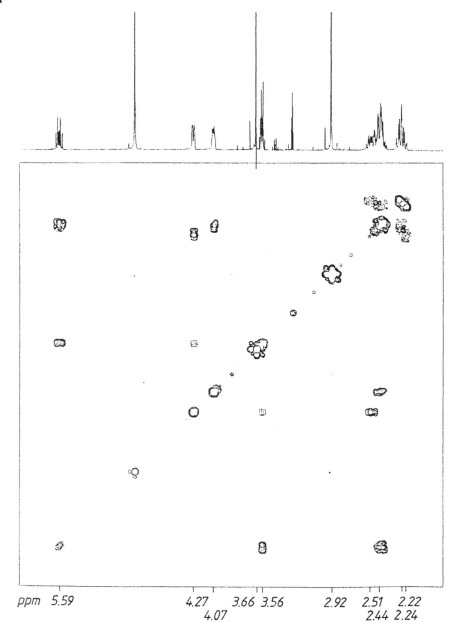

ppm   5.59   4.27   3.66 3.56   2.92   2.51   2.22
                                4.07   2.44 2.24

Problem 53, continued

(**b**) ^1H NMR spectrum with NOE difference spectrum, irradiation at $\delta_H = 2.92$; (**c**) ^{13}C NMR spectra, each with the ^1H broadband decoupled spectrum below and NOE enhanced coupled spectrum (gated decoupling) above;

53

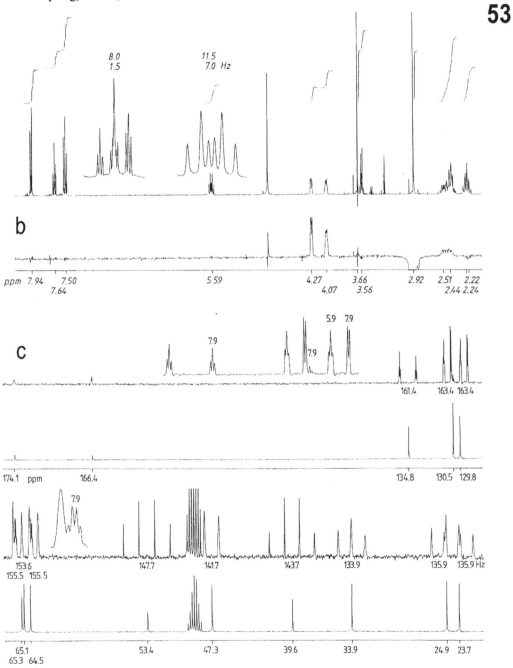

Problem 53, continued

(**d**) *CH* COSY and *CH* COLOC plots in one diagram with enlarged section (δ_C = 64.5 - 65.3 / δ_H = 2.44 - 3.56).

53d

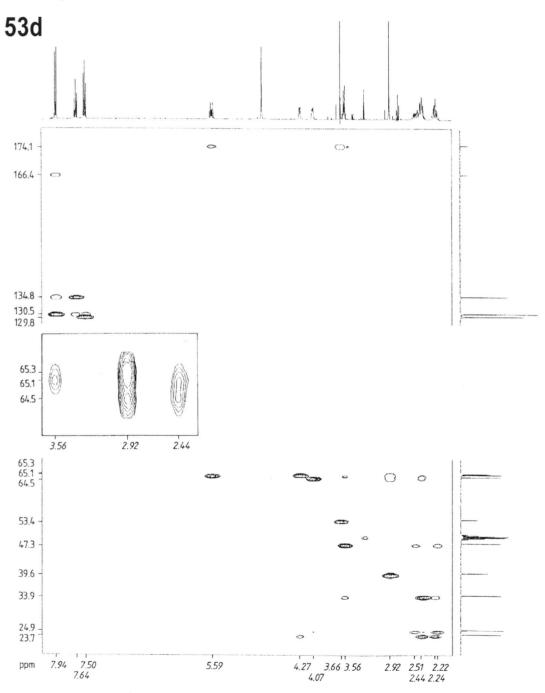

Problem 54

Amongst products isolated from *Heliotropium spathulatum* (Boraginaceae) were 9 mg of a new alkaloid which gave a positive Ehrlich reaction with *p*-dimethylaminobenzaldehyde [45]. The molecular formula determined by mass spectrometry is $C_{15}H_{25}NO_5$. What is the structure of the alkaloid given the set of NMR results **54**? Reference [31] is useful in providing the solution to this problem.

Conditions: CDCl₃, 9 mg per 0.3 ml, 25 °C, 400 MHz (¹H), 100 MHz (¹³C). (**a**) *HH* COSY plot;

54a

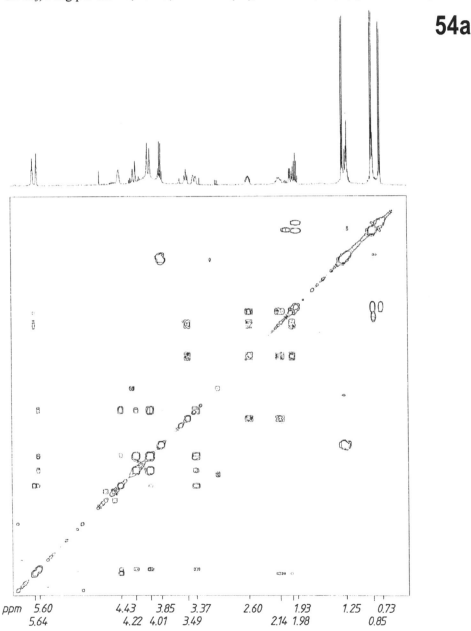

Problem 54, continued

(b) ^1H NMR partial spectrum beginning at $\delta_H = 1.98$ with NOE difference spectra, irradiations at the given chemical shifts;

54b

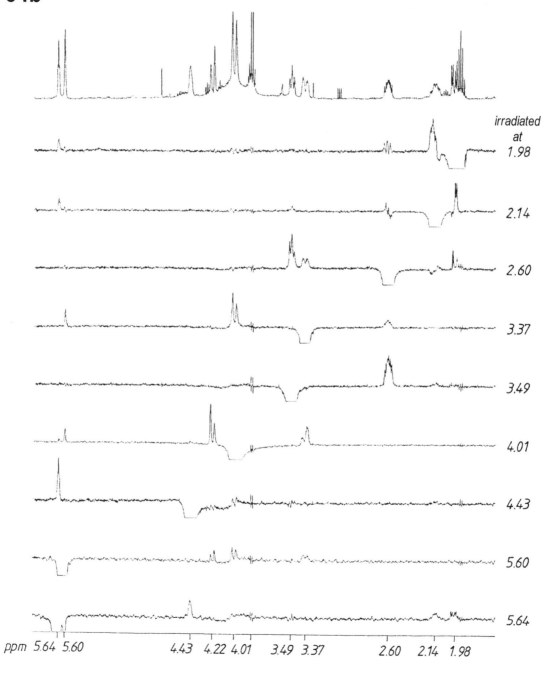

irradiated at

1.98

2.14

2.60

3.37

3.49

4.01

4.43

5.60

5.64

ppm 5.64 5.60   4.43 4.22 4.01   3.49 3.37   2.60   2.14 1.98

Problem 54, continued

(**c**) C*H* COSY plot with DEPT spectrum (C*H* and C*H₃* positive, C*H₂* negative);

54c

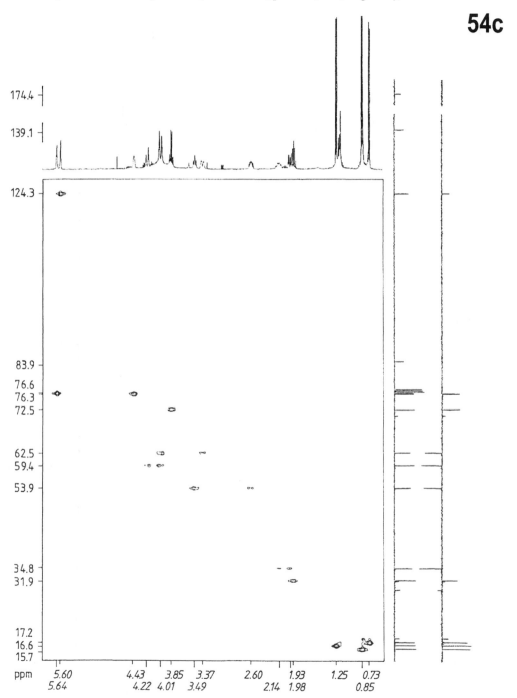

Problem 54, continued

(**d**) *CH* COLOC plot.

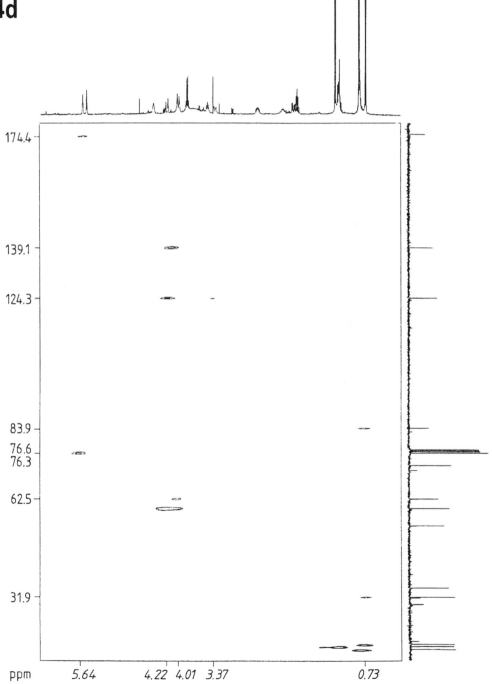

Problem 55

The NMR experiments **55** are obtained from actinomycin D in order to check the amino acid sequence, to assign proton-proton and some carbon-proton connectivities, and to deduce informations concerning proton distances and the spatial structure of both cyclopentapeptide lactone rings.
Conditions: CDCl$_3$, 10 mg per 0.3 ml, 25 °C, 500 MHz (^1H), 125 MHz (^{13}C). (**a**) *HH* COSY plot;

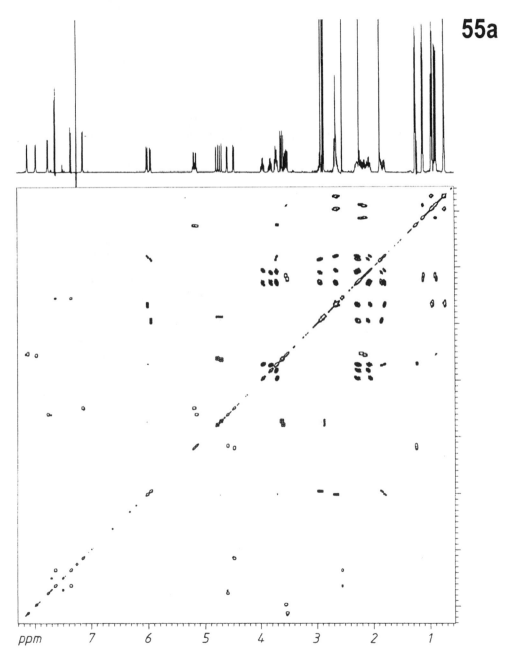

55a

Problem 55, continued

(**b**) *HH* TOCSY plot;

55b

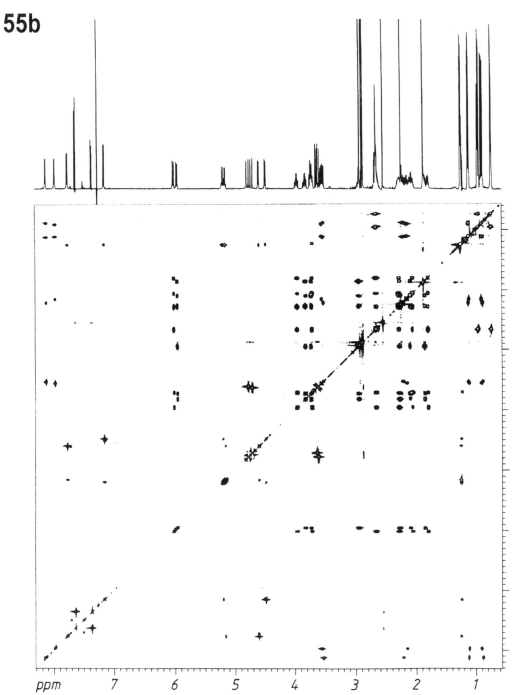

Problem 55, continued

(**c**) *HH* ROESY plot;

55c

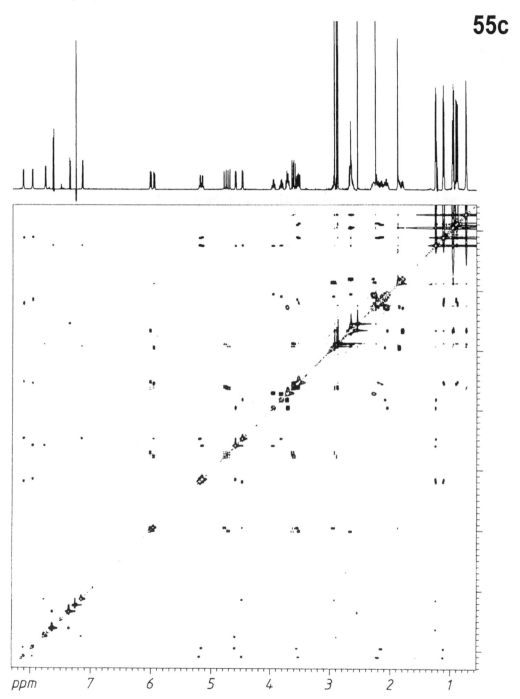

Problem 55, continued

(d) *H*C HMBC plot, section of heterocyclic carbons and carbonyl groups with ^{13}C chemical shifts
δ_C = 169.01, 168.52, 167.74, 167.65, 166.59, 166.56, 166.38, 166.18, 147.74, 145.94, 145.18, 140.53, 132.67, 130.33, 129.17, 127.66, 125.93;

55d

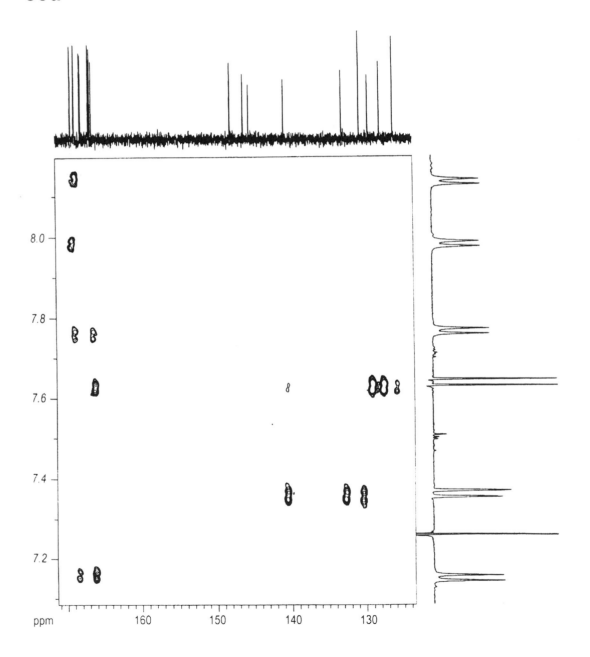

Problem 55, continued

(e) *H*C HMBC plot, section with connectivities of carbonyl carbon nuclei and amino protons.

55e

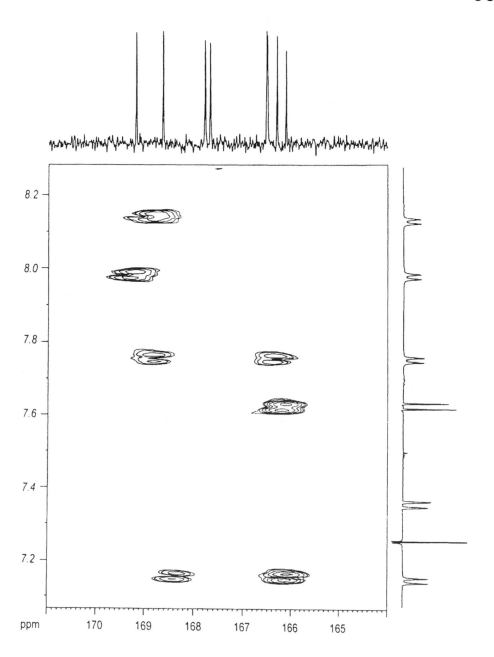

4   SOLUTIONS TO PROBLEMS

1   Dimethyl cis-cyclopropane-1,2-dicarboxylate

In the ^1H NMR spectrum it is possible to discern a four-spin system of the AMX_2 type for the three different kinds of proton of the cis-1,2-disubstituted cyclopropane ring. The X_2 protons form a doublet of doublets with cis coupling $^3J_{AX}$ (8.5 Hz) and trans coupling $^3J_{MX}$ (6.7 Hz). The signals of protons H^A and H^M also show geminal coupling $^2J_{AM}$ (5.1 Hz) and splitting into triplets (two H^X) of doublets. The trans isomer would show a four-spin system of the type $AA'BB'$ or $AA'XX'$ (according to the shift difference relative to the coupling constant) for the ring protons.

^1H chemical shifts δ_H

1.20 H   2.02
              H   H
1.62 H   CO$_2$CH$_3$
        CO$_2$CH$_3$ 3.64

Multiplicities and coupling constants (Hz)

                        dd
td H^A H^X H^X
td H^M CO$_2$CH$_3$
           CO$_2$CH$_3$

$^3J_{AX}$ = 8.5 (cis)
$^3J_{MX}$ = 6.7 (trans)
$^2J_{AM}$ = 5.1 (geminal)

2   Ethylacrylate

The empirical formula implies two double-bond equivalents. The ^1H NMR spectrum shows multiplet systems whose integral levels are consistent with the eight H atoms of the empirical formula in a ratio of 3:2:3. The triplet at δ_H = 1.3 and the quartet at δ_H = 4.2 with the common coupling constant of 7 Hz belong to the A_3X_2 system of an ethoxy group, $-OCH_2CH_3$. Three alkene protons between δ_H = 5.7 and 6.6 with the trans, cis and geminal couplings (13, 8 and 2 Hz) which are repeated in their coupling partners, identify the ABC system of a vinyl group, $-CH=CH_2$. If one combines both structural elements ($C_2H_5O + C_2H_3 = C_4H_8O$) and compares the result with the empirical formula ($C_5H_8O_2$), then C and O as missing atoms give a CO double bond in accordance with the second double-bond equivalent. Linking the structural elements together leads to ethyl acrylate.

^1H chemical shifts δ_H

         HB 6.10
5.75 HA   O
6.40 HC   O
              CH$_2$–CH$_3$ 1.30
                 4.20

Multiplicities and coupling constants (Hz)

         dd HB
dd HA   O
dd HC   O   q 7 Hz
              CH$_2$–CH$_3$
                 t 7 Hz

$^3J_{AB}$ = 8 (cis)
$^3J_{BC}$ = 13 (trans)
$^2J_{AC}$ = 2 (geminal)

Note how dramatically the roofing effects of the AB and BC part systems change the intensities of the doublet of doublets of proton B in spectrum 2.

3 *cis*-1-Methoxybut-1-en-3-yne

Three double-bond equivalents which follow from the empirical formula can be confirmed in the ^1H NMR spectrum using typical shifts and coupling constants. The proton signal at $\delta_H = 3.05$ indicates an ethynyl group with terminal proton H^A; an MX system in the alkene shift range with $\delta_M = 4.50$ and $\delta_X = 6.30$, respectively, and the coupling constants $^3J_{MX} = 8\ Hz$, reveal an ethene unit (–CH=CH–) with a *cis* configuration of the protons. The intense singlet at $\delta_H = 3.8$ belongs to a methoxy group, –OCH_3, whose –I effect deshields the H^X proton, whilst its +M effect shields the H^M proton. The bonding between the ethenyl and ethynyl groups is reflected in the long-range couplings $^4J_{AM} = 3$ and $^5J_{AX} = 1\ Hz$.

^1H chemical shifts δ_H

OCH₃ 3.80
6.30 HˣHₓ
HᴬHᴬ 3.05
4.50 Hᴹ

Multiplicities and coupling constants (Hz)

OCH₃ s
dd Hˣ
Hᴬ dd
dd Hᴹ

$^3J_{MX} = 8\ (cis)$
$^4J_{AM} = 3$
$^5J_{AX} = 1$

4 *trans*-3-(*N*-Methylpyrrol-2-yl)propenal

The empirical formula contains five double-bond equivalents. In the ^1H NMR spectrum a doublet signal at $\delta_H = 9.55$ stands out. This chemical shift value would fit an aldehyde function. Since the only oxygen atom in the empirical formula is thus assigned a place, the methyl signal at $\delta_H = 3.80$ does not belong to a methoxy group, but rather to an *N*-methyl group.

The coupling constant of the aldehyde doublet (*7.8 Hz*) is repeated in the doublet of doublets signal at $\delta_H = 6.3$. Its larger splitting of *15.6 Hz* is observed also in the doublet at $\delta_H = 7.3$ and indicates a CC double bond with a *trans* configuration of the *vicinal* protons.

The coupling of *7.8 Hz* in the signals at $\delta_H = 9.55$ and *6.3* identifies the (*E*)-propenal part structure. The large ^1H shift ($\delta_H = 9.55$) thus reflects the –M effect of the conjugated carbonyl group.

$\delta_H = 7.30$ Hᴹ   Hˣ 9.55

6.30 Hᴬ

Apart from the *N*-methyl group, three double-bond equivalents and three multiplets remain in the chemical shift range appropriate for electron rich heteroaromatics, $\delta_H = 6.2$ to *6.9*. *N*-Methylpyrrole is such a compound. Since in the multiplets at $\delta_H = 6.25$ and *6.80* the $^3J_{HH}$ coupling of *4.0* Hz is appropriate for pyrrole protons in the 3- and 4-positions, the pyrrole ring is deduced to be substituted in the 2-position.

^1H chemical shifts δ_H

3.80 H₃C 7.30 H   H 9.55
6.90 H
H 6.30
H 6.80
6.25 H

Multiplicities and coupling constants (Hz)

s H₃C   Hd   Hd
dd H
H dd
H dd
ddd H

$^3J_{1,2} = 7.8$
$^3J_{2,3} = 15.6$
$^5J_{3,4'} = 0.5$
$^3J_{3',4'} = 4.0$
$^3J_{3',5'} = 1.6$
$^3J_{4',5'} = 2.5$

5   1,9-Bis(pyrrol-2-yl)pyrromethene

The multiplets are sorted according to identical coupling constants which identify coupling partners, and after comparing *J*-values with characteristic pyrrole $^3J_{HH}$ couplings, two structural fragments **A** and **B** are deduced. The *ABX* system ($^3J_{AB}$ = 4.4, $^4J_{AX}$ = 2.2 and $^4J_{BX}$ =2.6 Hz for *X* = N*H*) with δ_H = 6.73 (*H^A*) and δ_H = 7.00 (*H^B*) belongs to the 2,5-disubstituted pyrrole ring **A**. The remaining three multiplets at δ_H = 6.35, 6.89 and 7.17 form an *ABC* system, in which each of the *vicinal* couplings of the pyrrole ring ($^3J_{AB}$ = 3.7 and $^3J_{AC}$ = 2.5 Hz) characterises a 2-monosubstituted pyrrole ring **B**.

¹H chemical shifts δ_H and multiplicities

$^3J_{AB}$ = 4.4
$^4J_{AX}$ = 2.2
$^4J_{BX}$ = 2.6

$^3J_{AB}$ = 3.7
$^3J_{AC}$ = 2.5
$^3J_{CX}$ = 2.5
$^4J_{CX}$ = 2.5
$^4J_{BX}$ = 2.5

Each of the two pyrrole rings occurs twice in the molecule judging by their integrated intensities relative to the methine singlet at δ_H = 6.80. Their connection with the methine group, which itself only occurs once (δ_H = 6.80), gives 1,9-bis(pyrrol-2-yl)pyrromethene **C**, a result which is illuminating in view of the reaction which has been carried out. An NOE between the protons at δ_H = 6.80 and 6.73 (not shown) confirmed the *cis* configuration of the central double bond as drawn. The symmetry is produced by exchange of the N*H* proton in the rings **A**.

6   3-Acetylpyridine

The ^1H NMR spectrum contains five signals with integral levels in the ratios 1 : 1 : 1 : 1 : 3; four lie in the shift range appropriate for aromatics or heteroaromatics and the fifth is evidently a methyl group. The large shift values (up to δ_H = 9.18, aromatics) and typical coupling constants (*8* and *5 Hz*) indicate a pyridine ring, which accounts for four out of the total five double-bond equivalents.

Four multiplets between δ_H = 7.46 and 9.18 indicate monosubstitution of the pyridine ring, either in the 2- or 3-position but not in the 4-position, since for a 4-substituted pyridine ring an *AA'XX'* system would occur. The position of the substituents follows from the coupling constants of the threefold doublet at δ_H = 7.46, whose shift is appropriate for a β-proton on the pyridine ring (**A**).

The *8 Hz* coupling indicates a proton in the γ-position (**B**); the *5 Hz* coupling locates a *vicinal* proton in position α (**C**), the additional *0.9 Hz* coupling locates the remaining proton in position α' (**D**) and thereby the β-position of the substituent.

This example shows how it is possible to pin-point the position of a substituent from the coupling constant of a clearly structured multiplet, whose shift can be established beyond doubt. The coupling constants are repeated in the multiplets of the coupling partners; from there the assignment of the remaining signals follows without difficulty. A monosubstituted pyridine ring and a methyl group add up to C_6H_7N. The atoms C and O which are missing from the empirical formula and one additional double-bond equivalent indicate a carbonyl group. The only structure compatible with the presence of these fragments is 3-acetylpyridine.

$$^3J_{4,5} = 8.0$$
$$^3J_{5,6} = 5.0$$
$$^4J_{2,4} = 1.8$$
$$^4J_{4,6} = 2.2$$
$$^5J_{2,5} = 0.9$$

7 Anthracene-1,8-dialdehyde

The ^1H NMR spectrum displays six signals with integral levels in the ratios $1 : 2 : 1 : 2 : 2 : 2$ in accordance with ten protons present in the empirical formula. All chemical shifts indicate protons attached to trigonal carbons. Three of those belong to benzenoid protons which are deshielded by electron withdrawing substituents ($\delta_A = 7.80$ dd, $\delta_M = 8.28$ and $\delta_X = 8.44$). The doublet of doublets at $\delta_H = 7.80$ displays two *ortho* couplings (8.5 and 7.0 Hz) so that the three-spin system (*AMX*) reveals a 1,2,3-trisubstituted benzenoid ring **A**. The electron withdrawing substituent turns out to be the aldehyde group **B** according to the chemical shift ($\delta_H = 10.40$). Partial structures **A** and **B** occur twice in the molecule on the basis of the integral levels of their proton signals in spectrum 7 giving rise to the partial formula $C_{14}H_8O_2$ which already includes both oxygen atoms, so that no phenolic OH is present. Thus, the remaining two singlet signals belong to two methine fragments **C** and **D**, each one representing formally one half of a double bond. Both complete the molecular formula $C_{16}H_{10}O_2$. One of the methine protons is extremely deshielded ($\delta_H = 11.05$).

The substructures **A** - **D** account for eleven of the total of twelve double bond equivalents so that the compound contains one additional ring as present in anthracene or phenanthrene skeletons.

From all symmetrically substituted anthracene and phenanthrene dialdehydes **E - H** containing 1,2,3-trisubstituted benzene rings,

only anthracene-1,8-dialdehyde **E** explains six proton signals (not five as expected for regioisomers **F-H**) in the ^1H NMR spectrum **7** as well as the outstanding deshielding of the methine proton ($\delta_H = 11.05$) attached to C-9 arising from accumulated anisotropic and mesomeric effects of two adjacent aldehyde groups.

8 *trans*-Stilbene-4-aldehyde

The singlet signal at $\delta_H = 10.0$ indicates an aldehyde function **A**. An *AB* two-spin system ($\delta_A = 7.16$ and $\delta_B = 7.27$) overlapping with the chloroform signal present in deuteriochloroform (99.5 % $CDCl_3$, 0.5 % $CHCl_3$) with a typical *trans* coupling of $^3J_{AB} = 16\ Hz$ reveals a CC double bond **B** with protons in *trans* configuration. An *AA'XX'*-four-spin system ($\delta_{AA'} = 7.67$ and $\delta_{XX'} = 7.89$, two protons for each signal according to the integral levels) indicates a *p*-disubstituted benzenoid ring **C**. The three remaining signals which include five protons according to the integral levels ($\delta_A = 7.32$, $\delta_{MM'} = 7.40$ and $\delta_{XX'} = 7.56$) belong to a monosubstituted benzene ring **D**.

Rational arrangement of these partial structures leads to *trans*-stilbene-4-aldehyde **E**.

The alternative *trans-p*-phenylcinnamic aldehyde **F** would display an additional $^3J_{HH}$ coupling between the aldehyde proton and the *vicinal* alkene proton of the double bond which is not observed in spectrum **8** (but in spectrum **4** for comparison).

Protons of substructure **B** and **C** are assigned by means of the mesomeric effect of the aldehyde group which deshields the protons in *o*-position of the attached *p*-disubstituted benzenoid ring and in β-position of the central CC double bond; *ortho*-protons of the monosubstituted benzenoid ring **D** split into a doublet because of one *ortho* coupling (*7.5 Hz*) while the *meta*-protons split into a triplet because of two *ortho* couplings.

9   6,4'-Dimethoxyisoflavone

Two intense signals at $\delta_H = 3.70$ and *3.80* identify two methoxy groups as substituents. The benzenoid proton resonances arise from two subspectra, an *AA'XX'* system ($\delta_{AA'} = 6.83$, $\delta_{XX'} = 7.37$) and an *ABX* system ($\delta_A = 7.11$, $\delta_B = 7.26$, $\delta_X = 7.51$). The *AA'XX'* part spectrum indicates a *para*-disubstituted benzene ring, and locates one methoxy group in the 4'-position of the phenyl ring **B**. A doublet of doublets with *ortho* and *meta* coupling (*9* and *3 Hz*, respectively) belongs to the *ABX* system, from which a 1,2,4-trisubstituted benzene ring (ring *A*) is derived. Hence the solution is isomer 6,4'- or 7,4'-dimethoxyisoflavone **A** or **B**.

The decisive clue is given by the large shift ($\delta_H = 7.51$) of the proton marked with an asterisk, which only shows one *meta* coupling. This shift value fits structure **A**, in which the –*M* effect and the anisotropy effect of the carbonyl group lead to deshielding of the proton in question. In **B** the +*M* effect of the two *ortho* oxygen atoms would lead to considerable shielding. The methoxy resonances cannot be assigned conclusively to specific methoxy groups in the presented spectrum.

10   Catechol (3,5,7,3',4'-pentahydroxyflavane)

Sesquiterpenes and flavonoids (flavones, flavanones, flavanes) are two classes of natural substances which occur frequently in plants and which have 15 C atoms in their framework. The nine double-bond equivalents which are contained in the empirical formula, ^1H signals in the region appropriate for shielded benzene ring protons (δ_H = 5.9 - 6.9) and phenolic OH protons (δ_H = 7.9 - 8.3) indicate a flavonoid.

In the ^1H NMR spectrum five protons can be exchanged by deuterium. Here the molecular formula permits only OH groups. The shift values (δ_H above 7.9) identify four phenolic OH groups and one less acidic alcoholic OH function (δ_H = 4.0, overlapping).

Between δ_H = 5.8 and 6.1 the ^1H signals appear with typical *ortho* and *meta* couplings. The small shift values show that the benzene rings are substituted by electron donors (OH groups). In this region two subspectra can be discerned: an *AB* system with a *meta* coupling (2.2 Hz) identifies a tetrasubstituted benzene ring **A** with *meta* H atoms; an *ABM* system with one *ortho* and one *meta* coupling (8.1 and 1.9 Hz, respectively) indicates a second benzene ring **B** with a 1,2,4-arrangement of the H atoms. Eight of the nine double-bond equivalents are thus assigned places.

Following the principle that the coupling partner will have the same coupling constant, one can identify in the aliphatic region a C$_3$ chain as a further part structure, **C**.

Assembly of fragments **A-C**, taking into account the ninth double-bond equivalent, leads to the 3,5,7,3',4'-pentahydroxyflavane skeleton **D** and to the following assignment of ^1H chemical shifts:

The relative configurations of phenyl ring **B** and the OH groups on ring **C** follows from the *antiperiplanar* coupling (8.3 Hz) of the proton at δ_H = 4.56. The coupling partner 3-H at δ_H = 4.0 shows this coupling a second time (pseudotriplet 't' with 8.3 Hz of doublets *d* with 5.0 Hz), because one of the neighbouring methylene protons is also located in a position which is *antiperiplanar* relative to 3-H (8.3 Hz) and another is located *syn* relative to 3-H (5.0 Hz). Hence one concludes that this compound is catechol or its enantiomer. The stereoformula **E** shows those coupling constants (Hz) which are of significance for deriving the relative configuration.

11 Methyloxirane and monordene

The relationship ($^3J_{cis} > {}^3J_{trans}$; cf. problem **1**), which applies to cyclopropane, also holds for the *vicinal* couplings of the oxirane protons (spectrum **11a**) with the exception that here values are smaller owing to the electronegative ring oxygen atoms. As spectrum **11a** shows, the *cis* coupling has a value of *3.9 Hz* whereas the *trans* coupling has a value of *2.6 Hz*. The proton at $\delta_H = 2.84$ is thus located *cis* relative to the proton at $\delta_H = 2.58$ and *trans* relative to the proton at $\delta_H = 2.28$. The coupling partners can be identified by their identical coupling constants where these can be read off precisely enough. Thus the methyl protons couple with the *vicinal* ring protons (*5.1 Hz*), the *cis* ring protons (*0.4 Hz*) and the *trans* ring protons (*0.5 Hz*); however, the small difference between these long-range couplings cannot be resolved in the methyl signal because of the large half-width, so that what one observes is a pseudotriplet.

Again following the principle that the same coupling constant holds for the coupling partner, the ^1H shift values of the protons on the positions C-1 to C-9 of monordene can be assigned (**A**, δ_H), as can the multiplicities and the coupling constants (**B**, *Hz*).

The *relative configurations* of *vicinal* protons follow from the characteristic values of their coupling constants. Thus *16.1 Hz* confirms the *trans* relationship of the protons on C-8 and C-9, *10.8 Hz* confirms the *cis* relationship of the protons on C-6 and C-7. The *2.0 Hz* coupling is common to the oxirane protons at $\delta_H = 3.00$ and *3.27*; this value fixes the *trans* relationship of the protons at C-4 and C-5 following a comparison with the corresponding coupling in the methyloxirane (*2.6 Hz*). The *anti* relationship of the protons 4-H and 3-H^A can be recognised from their *8.7 Hz* coupling in contrast to the *syn* relationship between 3-H^B and 4-*H* (*3.1 Hz*). Coupling constants which are almost equal in value (*3.2-3.7 Hz*) linking 2-*H* with the protons 3-H^A and 3-H^B indicate its *syn* relationship with these protons (3-H^A and 3-H^B straddle 2-*H*).

geminal couplings: Position 3 : $^2J_{AB}$ = 14.9 Hz
Position 11 : $^2J_{AB}$ = 16.3 Hz

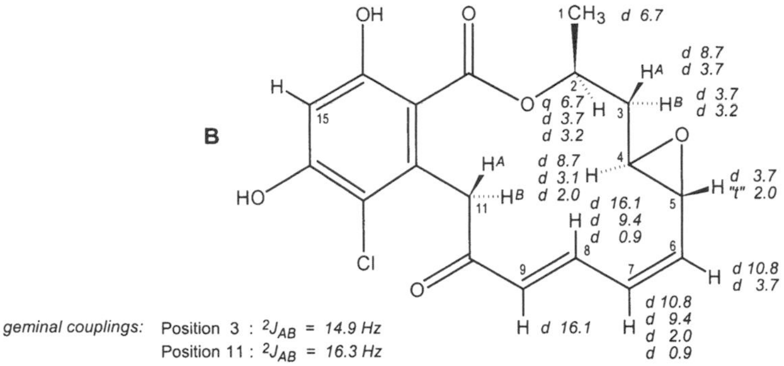

The absolute configuration, (2*S*,4*S*,5*S*) as shown or (2*R*,4*R*,5*R*), cannot be deduced by NMR. For larger structures the insertion of the shift values and the coupling constants in the stereo projection of the structural formula, from which one can construct a Dreiding model, proves useful in providing an overview of the stereochemical relationships.

12   2-Methyl-6-(*N*,*N*-dimethylamino)-*trans*-4-nitro-*trans*-5-phenyl-cyclohexene

An examination of the cross signals of the *HH* COSY diagram leads to the proton connectivities shown in **A** starting from the alkene proton at δ_H = *5.67*.

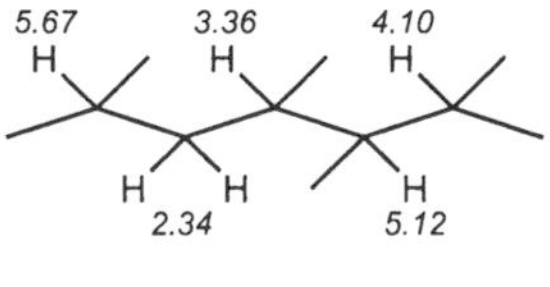

Strong cross signals linking the CH_2 group (δ_H = *2.34*) with the proton at δ_H = *3.36* confirm the regioselectivity of the Diels-Alder reaction and indicate the adduct **B**: the CH_2 is bonded to the phenyl-C*H* rather than to the nitro-C*H* group; if it were bonded to the latter, then cross signals for δ_H = *2.34* and *5.12* would be observed.

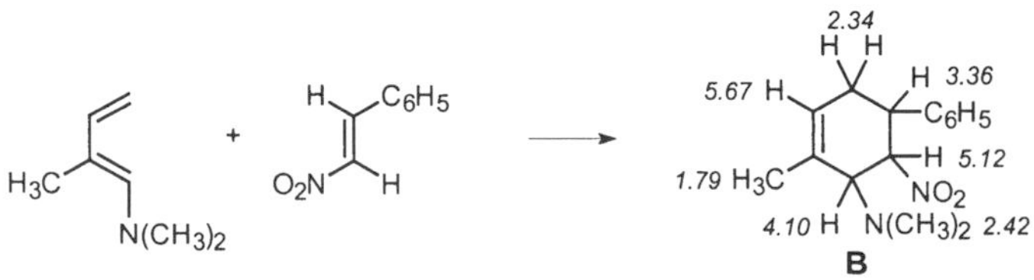

The relative configuration **C** is derived from the coupling constants of the ^1H NMR spectrum: the *11.9 Hz* coupling of the phenyl-C*H* proton (δ_H = *3.36*) proves its *antiperiplanar* relationship to the nitro-CH proton (δ_H = 5.12). In its doublet of doublets signal a second *antiperiplanar* coupling of 9.2 Hz appears in addition to the one already mentioned, which establishes the *anti* positon of the CH proton at δ_H = 4.10 in the positon α to the N,N-dimethylamino group.

*: couplings are not observed for the A_2 system of methylene protons

13   (*E*)-3-(*N*,*N*-Dimethylamino)acrolein

First the *trans* configuration of the C-2–C-3 double bond is derived from the large coupling constant ($^3J_{HH}$ = 13 Hz) of the protons at δ_H = 5.10 and 7.11, whereby the middle CH proton (δ_H = 5.10) appears as a doublet of doublets on account of the additional coupling (8.5 Hz) to the aldehyde proton.

The two methyl groups are not equivalent at 303 K (δ_H = 2.86 and 3.14); rotation about the CN bond is 'frozen', because this bond has partial π character as a result of the mesomeric (resonance) effects of the dimethylamino group (+*M*) and of the aldehyde function (–*M*), so that there are *cis* and *trans* methyl groups. Hence one can regard 3-(*N*,*N*-dimethylamino)acrolein as a vinylogue of dimethylformamide and formulate a vinylogous amide resonance.

At 318 K the methyl signals coalesce. The half-width Δv of the coalescence signal is approximately equal to the frequency separation of the methyl signals at 308 K; its value is $\Delta\delta_H$ = 3.14 – 2.86 = 0.28, which at 250 MHz corresponds to Δv = 70 Hz. The following exchange or rotation frequency of the N,N-dimethylamino group is calculated at the coalescence temperature:

$$k = (\pi / \sqrt{2}) \times \Delta v = 155.5 \text{ s}^{-1}$$

Finally from the logarithmic form of the Eyring equation, the free enthalpy of activation, ΔG, of rotation of the dimethylamino group at the coalescence temperature (318 K) can be calculated:

$$\Delta G_{318} = 19.134 \, T_c \, [\, 10.32 + \log (T_c\sqrt{2} \, / \, \pi\Delta v) \,] \times 10^{-3} \text{ kJmol}^{-1}$$
$$= 19.134 \times 318 \, [\, 10.32 + \log (318 \times 1.414 \, / \, 3.14 \times 70) \,] \times 10^{-3}$$
$$= 19.134 \times 318 \, (\, 10.32 + 0.311 \,) \times 10^{-3}$$
$$= 64.7 \text{ kJ mol}^{-1} \, (\, 15.45 \text{ kcal mol}^{-1} \,)$$

14   cis-1,2-Dimethylcyclohexane

The temperature dependence of the ^{13}C NMR spectrum is a result of cyclohexane ring inversion. At room temperature (298 K) four average signals are observed instead of the eight expected signals for the non-equivalent C atoms of cis-1,2-dimethylcyclohexane. Below –20 °C ring inversion occurs much more slowly and at –50 °C the eight expected signals of the conformers **I** and **II** appear.

The coalescence temperatures lie between 243 and 253 K and increase as the frequency difference between the coalescing signals in the 'frozen' state increases. Thus the coalescence temperature for the pairs of signals at δ_C = 35.2/33.3 lies between 238 and 243 K; owing to signal overlap the coalescence point cannot be detected precisely here. The methyl signals at δ_C = 20.5 and 11.5 have a larger frequency difference ($\Delta\delta_C$ = 9 or Δv = 900 Hz at 100 MHz) and so coalesce at 253 K, a fact which can be recognised from the plateau profile of the average signal (δ_C = 16.4). Since the frequency difference of this signal (900 Hz) in the 'frozen' state (223 K) may be measured more precisely than the width at half-height of the coalescing signal at 253 K, the exchange frequency k of the methyl groups is calculated from Δv:

$$k = (\pi /\sqrt{2}) \times 900 = 1998.6 \text{ s}^{-1}$$

The free enthalpy of activation, ΔG, of the ring inversion at 253 K is calculated from the logarithmic form of the Eyring equation:

$$
\begin{aligned}
\Delta G_{253} &= 19.134\, T_c\,[\,10.32 + \log (T_c\sqrt{2}\, /\, \pi\Delta v)\,]\times 10^{-3}\ \text{kJmol}^{-1}\\
&= 19.134 \times 253\,[\,10.32 + \log (253 \times 1.414\, /\, 3.14 \times 900\,)\,]\times 10^{-3}\\
&= 19.134 \times 253\,(\,10.32 - 1.9\,)\times 10^{-3}\\
&= 40.8 \text{ kJ mol}^{-1}\ (\,9.75 \text{ kcal mol}^{-1}\,)
\end{aligned}
$$

The assignment of resonances in Table 14.2 results from summation of substituent effects as listed in Table 14.1. The data refer to conformer **I**; for conformer **II** the C atoms pairs C-1/C-2, C-3/C-6, C-4/C-5 and C-7/C-8 change places.

Table 14.1. Prediction of ^{13}C chemical shift of cis-I,2-dimethylcyclohexane in the 'frozen' state, using the cyclohexane shift of δ_C = 27.6 and substituent effects (e.g. Ref. 6, p. 316)

C-1	C-2	C-3	C-4	C-5	C-6
27.6	27.6	27.6	27.6	27.6	27.6
+ 1.4 αa	+ 6.0 αe	+ 9.0 βe	+ 0.0 γe	– 6.4 γa	+ 5.4 βa
+ 9.0 βe	+ 5.4 βa	– 6.4 γa	– 0.1 δa	– 0.2 δe	+ 0.0 γe
– 3.4 $\alpha a \beta e$	– 2.9 $\alpha e \beta a$	– 0.8 $\beta a \gamma a$	+ 0.0 $\gamma e \delta e$	+ 0.0 $\gamma a \delta e$	+ 1.6 $\beta a \gamma e$
34.6	36.1	29.4	27.5	21.0	34.6

Table 14.2. Assignment of the ^{13}C resonances of *cis*-1,2-dimethylcyclohexane

	C-1	C-2	C-3	C-4	C-5	C-6	1-CH$_3$ ax.	2-CH$_3$ eq.
predicted δ_C	34.6	36.1	29.4	27.5	21.0	34.6	----	----
observed δ_C, CD$_2$Cl$_2$, 223 K	33.3	35.2	27.1	28.6	20.1	33.8	11.5	20.5
observed δ_C, CD$_2$Cl$_2$, 298 K	34.9	34.9	31.9	24.2	24.2	31.9	16.4	16.4

15   5-Ethynyl-2-methylpyridine

The ^{13}C NMR spectrum illustrates the connection between carbon hybridisation and ^{13}C shift on the one hand and J_{CH} coupling constants on the other.

The compound clearly contains a methyl group (δ_C = 24.4, quartet, J_{CH} = 127.5 Hz, sp^3) and an ethynyl group (δ_C = 80.4, doublet, J_{CH} = 252.7 Hz, sp; δ_C = 80.8, a doublet as a result of the coupling $^2J_{CH}$ = 47.0 Hz). Of the five signals in the sp^2 shift range, three belong to CH units and two to quaternary carbon atoms on the basis of the $^1J_{CH}$ splitting (three doublets, two singlets). The coupling constant J_{CH} = 182.5 Hz for the doublet centred at δ_C = 152.2 therefore indicates a disubstituted pyridine ring **A** with a CH unit in one α-position. It follows from the shift of the quaternary C atom that the methyl group occupies the other α-position (δ_C = 158.4, α-increment of a methyl group, about 9 ppm, on the α-C atom of a pyridine ring, approximately δ_C = 150); the shielding ethynyl group occupies a β-position, as can be seen from the small shift of the second quaternary C atom (δ_C = 116.4). From this, two regioisomers **B** and **C** appear possible.

The additional doublet splitting ($^3J_{CH}$ = 2.4 Hz) of the methyl quartet decides in favour of **C**; long range coupling in **B** ($^4J_{CH}$, $^5J_{CH}$) of the methyl C atom to H atoms of the ethynyl group and of the pyridine ring would not have been resolved in the spectrum. The long range quartet splitting of the pyridine CH signal at δ_C = 122.7 (C-3, $^3J_{CH}$ = 3.7 Hz) confirms the 2-position of the methyl group and thus locates the ethynyl group in the 5-position, as in **C**.

^{13}C chemical shifts δ_C

CH multiplicities , CH coupling constants (Hz) , coupling *protons*

C-2	S		d 10.5 (6-*H*)	d 4.3 (4-*H*)	d 2.4 (3-*H*)
C-3	D 163.6		q 3.7 (CH$_3$)	d 1.8 (4-*H*)	
C-4	D 165.0		d 5.5 (6-*H*)	d 1.8 (3-*H*)	
C-5	S		d 4.3 (3-*H*)	d 4.3 (6-*H*)	d 4.3 (β-*H*) ("q")
C-6	D 182.5		d 5.5 (4-*H*)	d 1.8 (β-*H*)	
2-CH$_3$	Q 127.5		d 2.4 (3-*H*)		
C-α	S		d 47.0 (β-*H*)		
C-β	D 252.7				

16   5-Hydroxy-3-methyl-1*H*-pyrazole

The compound referred to as 3-methylpyrazolone **A** ought to show a quartet and a triplet in the aliphatic region, the former for the ring CH_2 group. However, only a quartet is observed in the sp^3 shift range in hexadeuteriodimethyl sulphoxide whilst at $\delta_C = 89.2$ a doublet is found with $J_{CH} = 174.6$ Hz. An sp^2-hybridised C atom with two cooperating $+M$ effects fits the latter, the effect which OH and ring NH groups have in 5-hydroxy-3-methyl-1*H*-pyrazole **B**. The very strong shielding ($\delta_C = 89.2$) could not be explained by NH tautomer **C**, which would otherwise be equally viable; in this case only a $+M$ effect of the ring NH group would have any influence.

C-3	S	d 6.7 (4-*H*)	q 6.7 (CH₃) ("qui")
C-4	D 174.6	q 3.7 (CH₃)	
C-5	S	d 3.0 (4-*H*)	
CH₃	Q 128.1		

¹³C chemical shifts δ_C   CH multiplicities , CH coupling constants (Hz) , coupling *protons*

17   *o*-Hydroxyacetophenone

The compound contains five double-bond equivalents. In the ^{13}C NMR spectrum all eight C atoms of the molecular formula are apparent, as a CH_3 quartet ($\delta_C = 26.6$) four CH doublets ($\delta_C = 118$-136) and three singlets ($\delta_C = 120.0, 162.2, 204.9$) for three quaternary C atoms. The sum of these fragments ($CH_3 + C_4H_4 + C_3 = C_8H_7$) gives only seven H atoms which are bonded to C; since the molecular formula only contains oxygen as a heteroatom, the additional eighth H atom belongs to an OH group.

Since two quaternary atoms and four CH atoms appear in the ^{13}C NMR spectrum, the latter with a benzenoid $^3J_{CH}$ coupling constant of 7-9 Hz, this is a disubstituted benzene ring, and the C signal with $\delta_C = 162.2$ fits a phenoxy C atom. The keto carbonyl ($\delta_C = 204.9$) and methyl ($\delta_C = 26.6$) resonances therefore point to an acetyl group as the only meaningful second substituent. Accordingly, it must be either *o*- or *m*-hydroxyacetophenone **A** or **B**; the *para* isomer would show only four benzenoid ^{13}C signals because of the molecular symmetry.

It would be possible to decide between these two by means of substituent effects, but in this case a conclusive decision is reached using the $^3J_{CH}$ coupling: the C atom marked with an asterisk in **B** would show no $^3J_{CH}$ coupling, because the *meta* positions are substituted. In the coupled ^{13}C NMR spectrum, however, all of the CH signals are split with $^3J_{CH}$ couplings of 7-9 Hz. The $^3J_{CH}$ pseudotriplet splitting of the resonance at $\delta_C = 118.2$ argues in favour of **A**; the origin of the additional

$^3J_{CH}$ coupling of the C atom marked with an asterisk in **A** is the intramolecular hydrogen bonding proton. This coupling also permits straightforward assignment of the closely spaced signals at δ_C = 118.2 (C-3) and 119.2 (C-5).

^{13}C chemical shifts δ_C

		CH multiplicities , CH coupling constants (Hz) , coupling *protons*						
C-1	S		m					
C-2	S		m					
C-3	D	166.6	d	7.0	(5-*H*)	d 7.0 (O*H*)	("t")	
C-4	D	161.1	d	9.1	(6-*H*)			
C-5	D	165.4	d	7.9	(3-*H*)			
C-6	D	160.8	d	8.0	(4-*H*)			
C-α	S		q	5.5	(C*H*$_3$)	d 5.5 (6-*H*)	("qui")	
C-β	Q	128.1						

18   Potassium 1-acetonyl-2,4,6-trinitrocyclohexa-2,5-dienate

The ^{13}C NMR spectrum shows from the signals at δ_C = 205.6 (singlet), 47.0 (triplet) and 29.8 (quartet) that the acetonyl residue with the carbonyl group intact (δ_C = 205.6) is bonded to the trinitrophenyl ring. Only three of the four signals which are expected for the trinitrophenyl ring from the molecular symmetry (C-1, C-2,6, C-3,5, C-4) are found here (δ_C = 133.4, 127.6, 121.6); however, a further doublet signal (δ_C = 34.5 with J_{CH} = 145.6 Hz) appears in the aliphatic shift region. This shows that the benzene C*H* unit rehybridises from trigonal (sp^2) to tetrahedral (sp^3), so that a Meisenheimer salt **A** is produced.

Signal assignment is then no problem; the C atoms which are bonded to the nitro groups C-2,6 and C-4 are clearly distinguishable in the ^{13}C NMR spectrum by the intensities of their signals.

^{13}C chemical shifts δ_C

		CH multiplicities , CH coupling constants (Hz) , coupling *protons*						
C-1	D	145.6	t	4.5	(3,5-*H*$_2$)			
C-2,6	S		m					
C-3,5	D	166.2	d	4.4	(1-*H*)	d 4.4 (5/3-*H*)	("t")	
C-4	S		m					
C-α	T	130.1						
C-β	S		d	5.9	(1-*H*)	t 5.9 (α-*H*$_2$)	q 5.9 (γ-*H*$_3$)	("sep")
C-γ	Q	127.2						

19   *trans*-3-[4-(*N,N*-Dimethylamino)phenyl]-2-ethylpropenal

The relative configuration at the CC double bond can be derived from the $^3J_{CH}$ coupling of the aldehyde ^{13}C signal at $\delta_C = 195.5$ in the coupled ^{13}C NMR spectrum; as a result of this coupling, a doublet (with 11.0 Hz) of triplets (with 4.9 Hz) is observed. The 11.0 Hz coupling points to a *cis* configuration of aldehyde C and alkene *H*; the corresponding *trans* coupling would have a value of *ca* 15 Hz (reference substance: methacrolein, Table 2.11). The aldehyde and *p*-dimethylamino-phenyl groups therefore occupy *trans* positions.

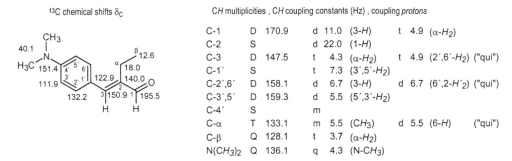

^{13}C chemical shifts δ_C

		CH multiplicities , CH coupling constants (Hz) , coupling *protons*						
C-1	D	170.9	d	11.0	(3-*H*)	t	4.9	(α-*H*$_2$)
C-2	S		d	22.0	(1-*H*)			
C-3	D	147.5	t	4.3	(α-*H*$_2$)	t	4.9	(2′,6′-*H*$_2$) ("qui")
C-1′	S		t	7.3	(3′,5′-*H*$_2$)			
C-2′,6′	D	158.1	d	6.7	(3-*H*)	d	6.7	(6′,2-*H*′$_2$) ("qui")
C-3′,5′	D	159.3	d	5.5	(5′,3′-*H*$_2$)			
C-4′	S		m					
C-α	T	133.1	m	5.5	(CH$_3$)	d	5.5	(6-*H*) ("qui")
C-β	Q	128.1	t	3.7	(α-*H*$_2$)			
N(CH$_3$)$_2$	Q	136.1	q	4.3	(N-CH$_3$)			

20   *N*-Butylsalicylaldimine

In the ^{13}C NMR spectrum the signal of the *O*-trimethylsilyl group is missing near $\delta_C = 0$. Instead there is a doublet ($^1J_{CH} = 159.3$ Hz) of quartets ($^3J_{CH} = 6.1$ Hz) at $\delta_C = 164.7$ for an imino C atom and a triplet of multiplets at $\delta_C = 59.0$. Its $^1J_{CH}$ coupling of 141.6 Hz points to an *N*-CH$_2$ unit as part of an *n*-butyl group with further signals that would fit this arrangement at $\delta_C = 33.0$, 20.3 and 13.7. The product of the reaction is therefore salicylaldehyde *N*-(*n*-butyl)imine. Assignment of the individual shifts for the hydrocarbon pair C-3/C-5, which are shielded by the hydroxy group as a *+M* substituent in the *ortho* and *para* position, respectively, is achieved by observing the possible long-range couplings: C-6 couples with 1.8 Hz to 7-*H*; C-3 is broadened as a result of coupling with the *H*-bonding O*H*. Both the latter and the *cis* coupling of C-α with 7-*H* (7.3 Hz) point to the *E*-configuration of *N*-butyl and phenyl relative to the imino double bond.

^{13}C chemical shifts δ_C

			CH multiplicities , CH coupling constants (Hz) , coupling *protons*								
C-1	S		d	5.5	(3-*H*)	d	5.5	(3-*H*)	d	5.5	(7-*H*) ("q")
C-2	S		d	7.9	(4-*H*)	d	7.9	(6-*H*)	d	7.9	(7-*H*) ("q")
C-3	D	160.5	d	6.7	(5-*H*)	d	b	(O*H*)			
C-4	D	159.9	d	8.5	(6-*H*)	d	1.2	(3/5-*H*)			
C-5	D	163.3	d	7.6	(3-*H*)						
C-6	D	157.9	d	7.9	(4-*H*)	d	1.8	(7-*H*)			
C-7	D	159.3	d	6.1	(6-*H*)	t	6.1	(α-*H*$_2$)			("q")
C-α	T	141.6	d	7.3	(7-*H*)	qui	3.4	(β-*H*$_2$, γ-*H*$_2$)			
C-β	T	127.6	m								
C-γ	T	127.0	m								
C-δ	Q	126.3	qui	3.1	(β-*H*$_2$, γ-*H*$_2$)						

21 Benzo[b]furan

All ^{13}C signals appear in the region appropriate for sp^2-hybridised C atoms; hence it could be an aromatic, a heteroaromatic or a polyene. If the matching doublet contour signals (· .) along the orthogonals of the INADEQUATE experiment are connected (**A**) then the result is eight CC bonds, six of which relate to the benzene ring. For example, one can begin with the signal at δ_C = 107.3 and deduce the hydrocarbon skeleton **A**.

The coupled ^{13}C NMR spectrum identifies the C atoms at δ_C = 145.7 and 155.9 as C*H* and C, respectively, whose as yet unattached bonds go to an electron-withdrawing heteroatom which causes the large shift values. The C*H* signals which are not benzenoid, at δ_C = 107.3 and 145.7, show remarkably large coupling constants (177.2 and 201.7 Hz, respectively) and long-range couplings (12.7 and 11.6 Hz). These data are consistent with a 2,3-disubstituted furan ring (Tables 2.6 and 2.7); benzo[b]furan **B** is therefore the result.

		CH multiplicities , *CH* coupling constants (Hz) , coupling *protons*		
C-2	D	201.7	d 11.6 (3-*H*)	
C-3	D	177.2	d 12.7 (2-*H*)	d 3.0 (4-*H*)
C-3a	S	m		
C-4	D	163.0	d 7.4 (6-*H*)	
C-5	D	162.3	d 8.9 (7-*H*)	
C-6	D	159.2	d 7.4 (4-*H*)	
C-7	D	160.0	d 6.7 (5-*H*)	d 1.5 (6-*H* / 3-*H*)
C-7a	S	m		

Additional CC correlation signals (δ_C = 145.7 to 122.0; 128.4 to 125.1; 122.0 to 112.1) are the result of $^3J_{CC}$ coupling and confirm the assignments given above.

22 3-Hydroxypropyl 2-ethylcyclohexa-1,3-diene-5-carboxylate

The cross signals in the INADEQUATE plot show the CC bonds for two part structures **A** and **B**. Taking the ^{13}C signal at δ_C = 174.1 as the starting point the hydrocarbon skeleton **A** and additional C$_3$ chain **B** result.

Part structure **A** is recognised to be a 2,5-disubstituted cyclohexa-1,3-diene on the basis of its chemical shift values. The ethyl group is one substituent, the other is a carboxy function judging by the chemical shift value of δ_C = 174.1. The C*H* multiplicities which follow from the DEPT subspectra, 2C, 4C*H*, 5C*H*$_2$ and C*H*$_3$, lead to the C*H* part formula C$_2$ + C$_4$H$_4$ + C$_5$H$_{10}$ + C*H*$_3$ = C$_{12}$H$_{17}$. Comparison with the given molecular formula, C$_{12}$H$_{18}$O$_3$, indicates an O*H* group. Since the C atoms at δ_C = 62.1 and 58.6 are linked to oxygen according to their shift values and according to the molecular formula, **A** and **B** can be added together to form 3-hydroxypropyl 2-ethylcyclohexa-1,3-diene-5-carboxylate **C**.

The assignment of C-α' and C-γ' is based on the larger deshielding of C-α' by the two β-C atoms (C-γ' and C=O).

23 Hex-3-yn-1-ol

All six of the C atoms found in the molecular formula appear in the ^{13}C NMR spectrum. Interpretation of the $^{1}J_{CH}$ multiplets gives one CH_3 group (δ_C = 14.4), three CH_2 groups (δ_C = 12.6, 23.2 and 61.6) and two quaternary C atoms (δ_C = 76.6 and 83.0). The addition of these CH fragments ($CH_3 + C_3H_6 + C_2$) produces C_6H_9; the additional H atom in the molecular formula therefore belongs to an OH group. This is a part of a primary alcohol function CH_2OH, because a ^{13}C shift of δ_C = 61.6 and the corresponding splitting (triplet, $^{1}J_{CH}$ = 144.0 Hz) reflect the $-I$ effect of an attached O atom. The long-range triplet splitting of the CH_2O signal (6.3 Hz) indicates a neighbouring CH_2 group. This hydroxyethyl partial structure **A** is evident in the ^{1}H NMR spectrum also, in which the coupling proton may be identified by the uniformity of its coupling constants.

Obviously the exchange frequency of the OH protons is small in comparison with the coupling constant (*4.9 Hz*), so coupling between the OH and CH_2 protons also causes additional splitting of the ^{1}H signals (δ_H = *3.58* and *4.72*).

The additional triplet splitting (*2.2 Hz*) of the CH_2 protons (at δ_H = *2.32*) is the result of long-range coupling to the third CH_2 group of the molecule, which can be recognised at δ_H = *2.13* by the same fine structure. The larger coupling constant (*7.6 Hz*) is repeated in the triplet at δ_H = *1.07*, so that an ethyl group is seen as a second structural fragment **B** in accordance with the further signals in the ^{13}C NMR spectrum (δ_C = 12.6, T 130.4 Hz, and δ_C = 14.4, Q 127.9 Hz, t 5.4 Hz).

The long-range coupling of *2.2 Hz* which appears in **A** and **B**, two quaternary C atoms in the ^{13}C NMR spectrum with appropriate shifts (δ_C = 76.6 and 83.0) and the two double-bond equivalents (molecular formula $C_6H_{10}O$) suggest that a CC triple bond links the two structural fragments. Hence the compound is identified as hex-3-yn-1-ol (**C**) in accordance with the coupling patterns.

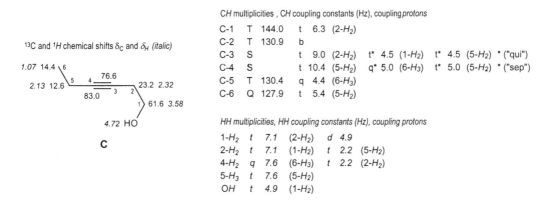

^{13}C and ^{1}H chemical shifts δ_C and δ_H (italic)

CH multiplicities , CH coupling constants (Hz), coupling protons

C-1 T 144.0   t   6.3 (2-*H₂*)
C-2 T 130.9   b
C-3 S   t   9.0 (2-*H₂*)   t* 4.5 (1-*H₂*)   t* 4.5 (5-*H₂*) * ("qui")
C-4 S   t 10.4 (5-*H₂*)   q* 5.0 (6-*H₃*)   t* 5.0 (5-*H₂*) * ("sep")
C-5 T 130.4   q   4.4 (6-*H₃*)
C-6 Q 127.9   t   5.4 (5-*H₂*)

HH multiplicities, HH coupling constants (Hz), coupling protons

1-*H₂* t 7.1 (2-*H₂*)   d 4.9
2-*H₂* t 7.1 (1-*H₂*)   t 2.2 (5-*H₂*)
4-*H₂* q 7.6 (6-*H₃*)   t 2.2 (2-*H₂*)
5-*H₃* t 7.6 (5-*H₂*)
OH   t 4.9 (1-*H₂*)

24   N,N-Diethylamino)ethyl 4-aminobenzoate hydrochloride (procaine hydrochloride)

The discussion centres on the two structural formulae **A** and **B**.

A   **B**

A choice can be made between these two with the help of published ^{13}C substituent effects [5,6] Z_i for the substituents ($-NH_2$, $-NH_3^+$, $-COOR$; see Section 2.5.4) on the benzene ring in **A** and **B**:

Substituent	Z_1	Z_o	Z_m	Z_p
$-NH_2$	18.2	-13.4	0.8	-10.0
$-NH_3^+$	0.1	-5.8	2.2	2.2
$-CO_2C_2H_5$	2.1	-1.0	-0.5	-3.9

Adding these substituent effects gives the following calculated shift values (as compared with the observed values in parentheses) for C-1 to C-4 of the *para*-disubstituted benzene ring in **A** and **B**:

	A	calc.		**B**	calc.	(observed)
δ_{C-1} = 128.5 + 2.1 + 2.2 =		132.8	δ_{C-1} = 128.5 + 2.1 $-$ 10.0 =		120.6	(115.5)
δ_{C-2} = 128.5 + 1.0 + 2.2 =		131.7	δ_{C-2} = 128.5 + 1.0 + 0.8 =		130.3	(131.3)
δ_{C-3} = 128.5 $-$ 0.5 $-$ 5.8 =		122.2	δ_{C-3} = 128.5 $-$ 0.5 $-$ 13.4 =		114.6	(113.1)
δ_{C-4} = 128.5 + 3.9 + 0.1 =		132.5	δ_{C-4} = 128.5 + 3.9 + 18.2 =		153.7	(153.7)

Substituent effects calculated for structure **B** lead to values which are not perfect but which agree more closely than for **A** with the measured ^{13}C shifts of the benzene ring carbon atoms. The diastereotopism of the NCH_2 protons in the ^1H NMR spectrum also points to **B** as the Newman projection **C** along the CH_2–ammonium-N bond shows:

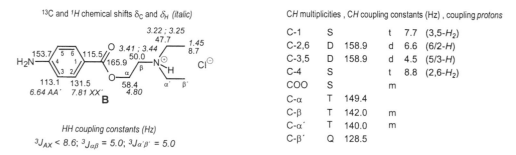

Hence one finds two overlapping pseudotriplets (δ_H = 3.41 and 3.44) for the NCH_2 group which appears only once and two overlapping quartets (δ_H = 3.22 and 3.25) for the NCH_2 groups which appear twice. Since the shift differences of the CH_2 protons are so small, the expected *AB* systems of the coupling partners approximate to A_2 systems at 200 MHz; thus one observes only the central multiplet signals of these *AB* systems.

The assignment of the ^{13}C NMR spectrum is based on the different $^1J_{CH}$ coupling constants of OCH_2 (149.4 Hz) and NCH_2 groups (140-142 Hz). With benzenoid $^3J_{CH}$ couplings the influence of the different electronegativities of the substituents on the coupling path (4.5 Hz for NH_2 and 6.6 Hz for CO_2R) and on the coupling C atom is very obvious (8.8 Hz for NH_2 at C-4 and 7.7 Hz for CO_2R at C-1).

^{13}C and ^1H chemical shifts δ_C and δ_H *(italic)*

HH coupling constants (Hz)
$^3J_{AX}$ < 8.6; $^3J_{\alpha\beta}$ = 5.0; $^3J_{\alpha'\beta'}$ = 5.0

CH multiplicities , CH coupling constants (Hz) , coupling *protons*

C-1	S		t	7.7 (3,5-H_2)
C-2,6	D	158.9	d	6.6 (6/2-*H*)
C-3,5	D	158.9	d	4.5 (5/3-*H*)
C-4	S		t	8.8 (2,6-H_2)
COO	S		m	
C-α	T	149.4		
C-β	T	142.0	m	
C-α'	T	140.0	m	
C-β'	Q	128.5		

25 5,5′-Bis-(hydroxymethyl)-2,2′-bifuran

The ^{13}C NMR spectrum displays five instead of ten carbon signals as expected from the empirical formula $C_{10}H_{10}O_4$. To conclude, two identical halves $C_5H_5O_2$ build up the molecular structure.

With oxygen as heteroatom in the empirical formula and according to the coupling constants of aromatic and heteroaromatic compounds listed in Table 2.5, the *AX* system with δ_A = 6.38, δ_X = 6.56 and J_{AX} = 3.5 Hz in the ^1H NMR spectrum reveals a 2,5-disubstituted furan ring **A**, matching well to $^3J_{3,4}$ = 3.4 Hz but neither to $^3J_{HH}$ = 1.8 Hz for 2,3- nor to $^4J_{HH}$ = 1.5 Hz for 3,4- nor to $^4J_{HH}$ = 0.9 Hz for 2,4-disubstitution. This agrees with two non-protonated carbons, both deshielded by an adjacent ring oxygen (δ_C = 155.4 and 145.2), and two *CH* carbon signals (δ_C = 109.2 and 106.1, doublets with one-bond couplings J_{CH} = 175.7 and 176.6 Hz, respectively), both shielded by the +*M* effect of the ring oxygen two bonds apart.

According to their integral levels (2:1), the remaining proton signals represent an A_2X system with $\delta_A = 4.41$, $\delta_X = 5.31$ and $J_{AX} = 6.0$ Hz. These protons arise from a hydroxymethylene group **B** because the ^{13}C NMR spectrum shows one CH_2 triplet with $\delta_C = 55.9$ but no additional CH doublet signal. $^3J_{HH}$ coupling (6.0 Hz) between CH_2 and OH protons is observed in this sample solution because solvation of the alcohol by hexadeuterioacetone prevents exchange by intermolecular hydrogen bonding.

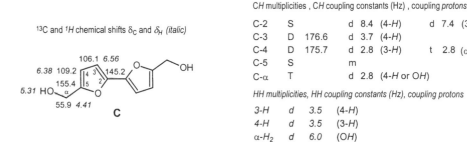

Partial structures **A** and **B** attached to each other represent one half ($C_5H_5O_2$) of the molecule according to the empirical formula ($C_{10}H_{10}O_4$). Therefore, the compound is 5,5'-dihydroxymethyl-2,2'-bifuran **C**. Carbon signals of the furan rings are assigned by means of two- and three-bond carbon-proton couplings as explained in the table. The proton signal of the furan ring proton $\delta_A = 6.38$) is slightly broadened by non-resolved allylic coupling ($^4J_{HH}$) to the methylene protons and, due to this, it is assigned.

^{13}C and ^1H chemical shifts δ_C and δ_H *(italic)*

CH multiplicities , CH coupling constants (Hz) , coupling *protons*						
C-2	S		d 8.4	(4-H)	d 7.4 (3-H)	("t")
C-3	D	176.6	d 3.7	(4-H)		
C-4	D	175.7	d 2.8	(3-H)	t 2.8 (α-H_2)	("q")
C-5	S		m			
C-α	T		d 2.8	(4-H or OH)		

HH multiplicities, HH coupling constants (Hz), coupling protons

3-H	d	3.5	(4-H)
4-H	d	3.5	(3-H)
α-H_2	d	6.0	(OH)
OH	t	6.0	(α-H_2)

26 *N*-Methyl-6,7-methylenedioxy-1-oxo-1,2,3,4-tetrahydroisoquinoline

The integral levels ($\sim 1 : 1 : 2 : 2 : 3 : 2$) of the proton signals include all eleven hydrogen atoms of the empirical formula $C_{11}H_{11}NO_3$ which implies seven double bond equivalents. An A_2X_2 system in the ^1H NMR spectrum ($\delta_A = 2.77$, $\delta_X = 3.49$ $^3J_{AX} = 7$ Hz) matching to two CH_2 carbon signals in the ^{13}C NMR spectrum ($\delta_C = 27.6$ and 47.8) indicates part structure **A**. A methyl signal in the ^1H as well as in the ^{13}C NMR spectrum ($\delta_H = 3.00$, three protons according to the integral level, $\delta_C = 34.8$, $J_{CH} = 137.3$ Hz) reveals an *N*-methyl group **B**. Two deshielded methylene protons with $\delta_H = 5.36$ and one methylene carbon with $\delta_C = 101.3$ (triplet with $J_{CH} = 174.1$ Hz) reveal a methylenedioxy fragment (formaldehyde acetal) fragment **C**. Two proton signals at $\delta_H = 6.47$ and 7.39 without resolved proton-proton couplings and six carbon signals between $\delta_C = 106.7$ and 150.0, two with, four without $^1J_{CH}$ couplings, reveal a tetrasubstituted benzene ring **D** with two protons in *p*-position. One remaining carbonyl signal with $\delta_C = 164.2$ belongs to a carboxylic acid derivative function **E** in conjugation to a CC double bond, referring to the shielding of this signal relative to non-conjugated carboxylic acid derivatives ($\delta_C > 170$).

34.8 3.00
—CH2—CH2—N
47.8 3.49

A

B

H2C 101.3 5.36

C

H 6.47
H 7.39

D

X
164.2
O

F

The carbon signal at $\delta_C = 133.4$ splits into a sextet with $^2J_{CH} \sim {}^3J_{CH} = 6$ Hz. This carbon of the benzenoid ring **D** is therefore attached to part structure **A**; $^3J_{CH}$ coupling with one benzenoid proton in *meta* position, $^2J_{CH}$ coupling with methylene protons at $\delta_A = 2.77$, $^3J_{CH}$ coupling with methylene protons at $\delta_X = 3.49$ explains this; the resulting doublet of triplets of triplets collapses to a pseudosextet ("sxt") due to equal coupling constants. Because only one nitrogen atom appears in the empirical formula, substructure **A**, **D** and **B** can be combined to an *N*-Methylphenylethylamine unit **G**. Substructures **A** - **F** found so far by NMR include five of the seven double bond equivalents. Two remaining double bond equivalents are thus attributed to two additional rings. One ring is closed by the open *ortho* positions of the phenyl ring in **G** and the methylenedioxy function **C**. The other arises from connection of the carboxy carbonyl with the remaining open bonds on the other side of the molecule, completing it to *N*-methyl-6,7-methylenedioxy-1-oxo-1,2,3,4-tetra-hydroisoquinoline **H**.

C   **G**   **F**   **H**

All carbon signals and resolved couplings can be assigned; C-6 is more deshielded ($\delta_C = 150.0$) than C-7 ($\delta_C = 146.5$) due to electron withdrawal of the carboxy carbonyl group in *para* position.

^{13}C and ^{1}H chemical shifts δ_C and δ_H (italic)

H

CH multiplicities , CH coupling constants (Hz) , coupling *protons*

C-1	S		m					
C-3	T	139.1	t 4.0 (4-H_2)					
C-4	T	129.3	t 3.5 (3-H_2)	d 3.5 (5-H)				
C-4a	S		t 6.0 (4-H_2)	t 6.0 (3-H_2)	d 6.0 (8-H)("sxt")			
C-5	D	164.4	t 3.0 (4-H_2)					
C-6	S		d 7.1 (8-H)					
C-7	S		d 6.3 (5-H)					
C-8	D	162.0						
C-8a	S		d 5.0 (5-H)	t 5.0 (4-H_2) ("q")				
OCH2O	T	174.1						
NCH3	Q	137.3						

27 Ethoxycarbonyl-4-(3-hydroxypropyl)-1-methylpyrrole

Here it is possible to consider how the starting materials may react and to check the result with the help of the spectra. Another approach would start by tabulating the ^{13}C shifts, CH multiplicities and CH coupling constants and where possible the ^1H shifts and the HH coupling constants (Table 27.1). From this it is possible to identify those parts of the starting materials that have remained intact and those which have been lost and also those H atoms which are linked to carbon and to heteroatoms.

Table 27.1. Interpretation of the NMR spectra in **27**

No.	δ_C	CH multiplicity	assignment	J_{CH} (Hz)	δ_H	HH multiplicity	J_{HH} (Hz)
1	161.2	S	CO(O)				
2	127.5	D	CH	181.6	6.73	d	1.9
3	122.8	S	C				
4	121.9	S	C				
5	117.0	D	CH	172.2	6.52	d	1.9
6	61.7	T	OCH$_2$	140.7	3.57	t	7.0
7	59.5	T	OCH$_2$	147.1	4.18	q	7.2
8	36.3	Q	NCH$_3$	140.2	3.78	s	
9	33.5	T	CH$_2$	126.4	1.74	qui	6.9
10	22.5	T	CH$_2$	126.3	2.43	t	7.0
11	14.2	Q	CH$_3$	126.7	1.26	t	7.2
CH partial formula			$C_{11}H_{16}(NO_3)$				

This evaluation reveals that the three substructures of the reagents that are also present in the product include the N-methyl group (signals 8), the ethoxycarbonyl group (signals 1, 7, 11) and the n-propyloxy group of the dihydro-$2H$-pyran ring (signals 6, 9, 10). The ethyl ester OCH$_2$ group can also be identified in the ^{13}C NMR spectrum because of its long-range quartet splitting (4.5 Hz). The H atom missing in the CH balance but present in the molecular formula appears in the ^1H NMR spectrum as a broad D$_2$O-exchangeable signal ($\delta_H = 3.03$); since the compound only contains one nitrogen atom in the form of an NCH$_3$ group, the signal at ($\delta_H = 3.03$) must belong to an OH group. Hence the dihydro-$2H$-pyran ring has opened.

By contrast, the aldehyde signals of the reagent **1** are missing from the NMR spectra. Instead an AB system appears in the ^1H NMR spectrum ($\delta_A = 6.52$, $\delta_B = 6.73$ with $J_{AB} = 1.9$ Hz) whilst in the ^{13}C NMR spectrum two doublets appear ($\delta_C = 117.0$ and 127.5) as well as two singlets ($\delta_C = 121.9$ and 122.8), of which one doublet ($\delta_C = 127.5$) is notable for the fact that it has a large CH coupling constant ($J_{CH} = 181.6$ Hz). This value fits the α-C atom of an enamine fragment (for the α-C of an enol ether fragment $J_{CH} \geq 190$ Hz would be expected). This leads to a 1,2,4-trisubstituted pyrrole ring **5**, given the three double-bond equivalents (the fourth has already been assigned to the carboxy group), the AB system in the ^1H NMR spectrum ($\delta_A = 6.52$, $\delta_B = 6.73$), the N-methyl group (signal 8) and the four ^{13}C signals in the sp^2 shift range ($\delta_C = 117$-127.5). The formation of the ring from reagents **1** and **2** via intermediates **3** and **4** can be inferred with no difficulty. All ^1H and ^{13}C signals can be identified without further experiment by using their shift values, multiplicities and coupling constants.

^{13}C and ^{1}H chemical shifts δ_C and δ_H (italic)

5

HH coupling constants (Hz)

$^{3}J_{3,5} = 1.9$; $^{3}J_{\alpha,\beta} = 7.0$; $^{3}J_{\beta,\gamma} = 7.0$; $^{3}J_{\alpha',\beta'} = 7.2$

CH multiplicities , CH coupling constants (Hz) , coupling protons

C-2	S		d	6.5	(3-H)	d	6.5 (5-H)	("t")
C-3	D	172.2	d	5.0	(5-H)	t	5.0 (6-H$_2$)	("q")
C-4	S		m					
C-5	D	181.6	d	7.0	(3-H	t	3.9 (6-H$_2$)	
NCH$_3$	Q	140.2	s					
COO	S		b					
C-α	T	126.3	m					
C-β	T	126.4	m					
C-γ	T	140.7	t	4.4	(β-H$_2$)	t	4.4 (α-H$_2$)	("qui")
C-α'	T	147.1	q	4.5	(β'-H$_2$)			
C-β'	Q	126.7	b					

28   p-Tolylsulphonyl-5-propylpyridine

The NMR spectra show that the product of the reaction contains:

■   the propyl group **A** of 1-ethoxy-2-propylbuta-1,3-diene,

■   the p-tolyl residue **B** from p-toluenesulphonyl cyanide.

■   and (on the basis of their typical shift values and coupling constants, e.g. $J_{CH} = 180.2$ Hz at $\delta_C = 150.5$, a disubstituted pyridine ring **C** (three ^{1}H signals in the ^{1}H NMR, three CH doublets in the ^{13}C NMR spectrum) with substituents in the 2- and 5-positions, because in the ^{1}H NMR spectrum the 8.2 Hz coupling ($^{3}J_{AM} = {}^{3}J_{3-H, 4-H}$) appears instead of the 5 Hz coupling (Table 2.5).

However, the ethoxy group of l-ethoxy-2-propylbuta-l,3-diene is no longer present. Evidently the p-toluenesulphonyl cyanide (2) undergoes [4+2] cycloaddition to l-ethoxy-2-propylbuta-1,3-diene (1). The resulting dihydropyridine 3 aromatises with 1,4-elimination of ethanol to form 2-p-tolyl-sulphonyl-5-propylpyridine (4).

1   **2**   **3**   **4**

Complete assignment is possible without further experiments using the characteristic shifts, multiplicities and coupling constants.

^{13}C and ^{1}H chemical shifts δ_C and δ_H *(italic)*

HH coupling constants (Hz)

$^{3}J_{AM} = 8.2$; $^{4}J_{MX} = 1.8$; $^{3}J_{AX} = {}^{3}J_{A'X'} < 8.4$;

$^{3}J_{\alpha,\beta} = {}^{3}J_{\beta\gamma} = 7.3$

CH multiplicities , CH coupling constants (Hz) , coupling protons

C-2	S		d 11.8 (6-*H*)	d 8.9 (4-*H*)			
C-3	D	170.3					
C-4	D	163.4	d 4.9 (6-*H*)	t 4.9 (α-H_2)			("q")
C-5	S		d 10.0 (6-*H*)	d 5.0 (3-*H*)	t 5.0 (α-H_2)	t 5.0 (β-H_2)	(d "qui")
C-6	D	180.2	d 5.9 (4-*H*)	t 5.9 (α-H_2)			("q")
C-α	T	127.0	b				
C-β	T	127.0	t 4.9 (α-H_2)	q 4.9 (γ-H_3)			("sxt")
C-γ	Q	126.2	t 3.9 (β-H_2)				
C-1'	S		t 8.9 (3',5'-H_2)				
C-2',6'	D	147.1	d 6.0 (6'/2'-*H*)				
C-3',5'	D	126.7	d 6.0 (2',6'-H_2)	q 6.0 (CH_3)			("qui")
C-4'	S		m				
4-CH_3	Q	127.0	t 4.4 (3',5'-H_2)				

29   6-Methoxytetralin-1-one

Almost all parts of the structure of this compound are already apparent in the ^{1}H NMR spectrum. It is possible to recognise:

- three methylene groups linked to one another, **A**,

- a methoxy group **B**,

- and a 1,2,4-trisubstituted benzene ring **C**, in the following way:

The signal at $\delta_H = 6.79$ splits into a doublet of doublets. The larger coupling (*8.7 Hz*) indicates a proton in the *ortho* position, the smaller (*2.5 Hz*) a further proton in a *meta* position, and in such a way that the *ortho* proton ($\delta_H = 7.97$) does not show any additional *ortho* coupling.

A   **B**   **C**

The ^{13}C NMR spectrum confirms

- three methylene groups **A** (δ_C = 23.6, 30.3, 39.1, triplets),
- the methoxy group **B** (δ_C = 55.7, quartet),
- the trisubstituted benzene ring **C** (three C*H* doublets and three non-protonated C atoms between δ_C = 113.3 and 164.6)
- and identifies additionally a keto-carbonyl group **D** at δ_C = 197.8.

Five double-bond equivalents can be recognised from the shift values (four for the benzene ring and one for the carbonyl group). The sixth double-bond equivalent implied by the molecular formula belongs to another ring, so that the following pieces can be drawn for the molecular jigsaw puzzle:

| B | C | D | A |

The methoxy group is a *+M* substituent, and so shields *ortho* protons and C atoms in *ortho* positions; the protons at δ_H = *6.67* and *6.79* reflect this shielding. The carbonyl group as a *–M* substituent deshields *ortho* protons, and is *ortho* to the proton at δ_H = *7.97*. With the additional double-bond equivalent for a ring, 6-methoxytetralin-1-one (**E**) results.

The difference between 2-C*H*$_2$ and 4-C*H*$_2$ is shown by the nuclear Overhauser enhancement (NOE) on the proton at δ_H = *6.67*, if the methylene protons at δ_H = *2.87* are irradiated. The assignment of the methylene C atoms can be read from the C*H* COSY segment. The C atoms which are in close proximity to one another at δ_C = 113.3 and 113.8 belong to C-5 and C-7. Carbon atom C-5 is distinguished from C-7 by the pseudo-quartet splitting ($^3J_{CH}$ = 3.4 Hz to 7-*H* and 4-*H*$_2$) that involves the methylene group in the *ortho* position.

^{13}C and ^{1}H chemical shifts δ_C and δ_H *(italic)*

E

CH multiplicities , CH coupling constants (Hz) , coupling *protons*

C-1	S		d	8.0	(8-*H*)	t	4.0	(2-*H*$_2$)	t	4.0	(3-*H*$_2$)	("qui")
C-2	T	127.3	b									
C-3	T	129.0	t	3.4	(2-*H*$_2$)	t	3.4	(4-*H*$_2$)				("qui")
C-4	T	130.0	b									
C-4a	S		d	4.0	(8-*H*)	t	4.0	(4-*H*$_2$)				("q")
C-5	D	158.3	d	3.4	(3-*H*)	t	3.4	(4-*H*$_2$)				("q")
C-6	S		m									
C-7	D	156.6	d	5.2	(6-*H*)							
C-8	D	161.2	s									
C-8a	S		m									
OCH$_3$	Q	144.5	s									

HH multiplicities, HH coupling constants (Hz), coupling *protons*

2-*H*$_2$	t	6.2	(3-*H*$_2$)			
3-*H*$_2$	"qui"	6.2	(2-*H*$_2$, 4-*H*$_2$)			
4-*H*$_2$	t	6.2	(3-*H*$_2$)			
5-*H*	d	2.5	(7-*H*)			
7-*H*	d	8.7	(8-*H*)	d	2.5	(5-*H*)
8-*H*	d	8.7	(7-*H*)			
OCH$_3$	s					

30 Triazolo[1,5-a]pyrimidine

Without comparative data on authentic samples, ^1H and ^{13}C NMR allow no differentiation between isomers **3** and **4**; chemical shifts and coupling constants are consistent with either isomer.

^{13}C and ^1H chemical shifts δ_C and δ_H *(italic)*

CH multiplicities , CH coupling constants (Hz) , coupling *protons*

C-2	D	208.1					
C-3a	S		m				
C-5	D	186.2	d	6.7	(7-H)	d 3.0 (6-H)	
C-6	D	174.6	d	9.1	(5-H)	d 3.0 (7-H)	
C-7	D	192.5	d	6.1	(5-H)	d 4.9 (6-H)	

HH coupling constants (Hz) of **3** and **4**
$^3J_{AM} = 4.3$; $^4J_{MX} = 2.0$; $^3J_{AX} = 6.8$

C-3	D	208.1					
C-5	D	192.4	d	6.1	(7-H)	d 4.9 (6-H)	
C6	D	174.6	d	9.1	(7-H)	d 3.0 (5-H)	
C-7	D	186.2	d	6.7	(5-H)	d 3.0 (6-H)	
C-8a	S		m			m	

However, in ^{15}N NMR spectra, the $^2J_{NH}$ coupling constants (≥ 10 Hz) are valuable criteria for structure determination. The ^{15}N NMR spectrum shows $^2J_{NH}$ doublets with 11.8, 12.8 and 15.7 Hz for all of the imino N atoms. Therefore, triazolo[1,5-a]pyrimidine (**3**) is present; for the [4,3-a] isomer **4**, nitrogen atom N-1 would appear as a singlet signal because it has no H atoms at a distance of two bonds. This assignment of the ^{15}N shifts is supported by a comparison with the spectra of derivatives which are substituted in positions 2 and 6 [8]. If a substituent is in position 6 then the 1.5 Hz coupling is lost for N-4; for substitution in position 2 or 6 a doublet instead of a triplet is observed for N-8. The ^{15}N shift and the $^2J_{NH}$ coupling constants of N-1 are considerably larger than for N-3 as a result of the electronegativity of the neighbouring N-8.

N-1	d	15.7	(2-H)			
N-3	d	12.8	(2-H)			
N-4	d	11.8	(5-H)			
N-8	d	5.7	(2-H)	d 5.7 (6-H)	("t")	

31 6-*n*-Butyltetrazolo[1,5-a]pyrimidine and 2-azido-5-*n*-butylpyrimidine

Tetrazolo[1,5-a]pyrimidine (**1**) exists in equilibrium with its valence isomer 2-azidopyrimidine (**2**).

In all types of NMR spectra (^1H, ^{13}C, ^{15}N), 2-azidopyrimidine (**2**) can be distinguished by the symmetry of its pyrimidine ring (chemical equivalence of 4-*H* and 6-*H*, C-4 and C-6, N-1 and N-3) from tetrazolo[1,5-a]pyrimidine (**1**) because the number of signals is reduced by one. Hence the prediction in Table 30.1 can be made about the number of resonances for the *n*-butyl derivative.

Table 30.1. The number of signals from **1** and **2** in the NMR spectra

Compound	^1H signals	Number of ^{13}C signals	^{15}N signals
1	6	8	5
2	5	7	4

All of the NMR spectra indicate the predominance of the tetrazolo[1,5-*a*]pyrimidine **1** in the equilibrium by the larger intensity (larger integral) of almost all signals, although the non-equivalence of the outer *n*-butyl C atoms in both isomers (at δ_C = 22.5 and 13.9) cannot be resolved in the ^{13}C NMR spectrum. By measuring the integrated intensities, for example, one obtains for the signals showing $^2J_{NH}$ splitting (of 12.0 and 11.5 Hz, respectively) recognisable signal pairs of the pyrimidine N atoms (**1**, at δ_N = 275.6; **2**, at δ_N = 267.9) of integrated intensities in the ratio 23 to 8.4 (mm). Since two N nuclei generate the signal at δ_N = 267.9 because of the chemical equivalence of the ring N atoms in **2** its integral must be halved (4.2). Thus we obtain

$$\% \, 2 = 100 \times 4.2 \, / \, (23 + 4.2) = 15.4 \, \% \, .$$

The evaluation of other pairs of signals in the ^1H and ^{15}N NMR spectra leads to a mean value of 15.7 ± 0.5 % for **2**. Therefore, 6-*n*-butyltetrazolo[1,5-*a*]pyrimidine (**1**) predominates in the equilibrium with 84.3 ± 0.5 %.

Assignment of the signals is completed in Table 30.2. The criteria for assignment are the shift values (resonance effects on the electron density on C and N), multiplicities and coupling constants. Because the difference between them is so small, the assignment of N-8 and N-9 is interchangeable.

Table 30.2. Assignment of the signals from 6-*n*-butyl tetrazolo[1,5-*a*]pyrimidine (**1**) and 2-azido-5-*n*-butylpyrimidine (**2**)

^{13}C, ^1H and ^{15}N chemical shifts δ_C, δ_H (italic) and δ_N (bold)	multiplicities, coupling constants (Hz), coupling *protons*				
	Position	J_{CH}	$^3J_{CH}$	$^3J_{HH}$	$^2J_{NH}$
δ 13.9 *1.00*	C-3a	S	d 14.7 (5-H)		
γ 22.5 *1.55*	N-4	---------		---------	d 12.0 (5-H)
β 29.6 *1.95*	C-5	D 185.9	d 5.0 (7-H) t 5.0 (α-H2)	d 1.8 (7-H)	
131.9 *9.36*	C-6	S	b		
α 32.7	C-7	D 193.1	d 5.0 (5-H) t 5.0 (α-H2)	d 1.8 (5-H)	
3.24 236.6 **N** 347.0	C-α	T 128.5	b	t 7.4 (β-H2)	
129.0 **N** 402.3	C-β	T 128.7	b	qui 7.4 (α,γ-H4)	
9.83 161.9 **275.6 N** 154.7 **N** 310.0	C-γ	T 126.6	b	sxt 7.4 (β,δ-H5)	
1	C-δ	Q 125.5	b	t 7.4 (γ-H2)	
δ 13.9 *0.95*	N-1,3	---------		---------	d 11.5 (4/6-H)
γ 22.5 *1.45*	C-2	S	t 12.5 (4,6-H2)		
β 29.5 *1.75*	C-4,6	D 180.2	d 5.0 (6/4-H) t 5.0 (α-H2)		
2.80 33.3 α	C-5	S	b		
132.2 **N** 238.0	C-α	T 127.7	b	t 7.4 (β-H2)	
8.67 159.4 **N** 236.7	C-β	T overlapping	b	qui 7.4 (α,γ-H4)	
267.9 N 160.2 **N** 109.0	C-γ	T overlapping	b	sxt 7.4 (β,δ-H5)	
2	C-δ	Q overlapping	b	t 7.4 (γ-H2)	

32 Hydroxyphthalide

The ^1H NMR spectrum does not show a signal for either a carboxylic acid or an aldehyde function. Instead, a D_2O-exchangeable signal indicates a less acidic OH proton ($\delta_H = 4.8$) and a non-exchangeable signal appears at $\delta_H = 6.65$. The latter fits a CH fragment of an acetal or hemiacetal function which is strongly deshielded by two O atoms, also confirmed by a doublet at ($\delta_C = 98.4$ with $J_{CH} = 174.6$ Hz in the ^{13}C NMR spectrum. According to this it is not phthalaldehydic acid (*1*) but its acylal, hydroxyphthalide (*2*) .

The conclusive assignment of the ^1H and ^{13}C signals of the *ortho*-disubstituted benzene ring at 80 and 20 MHz, respectively, encounters difficulties. However, the frequency dispersion is so good at 400 and 100 MHz, respectively, that the *HH* COSY in combination with the *CH* COSY diagram allows a conclusive assignment to be made. Proton connectivities are derived from the *HH* COSY; the *CH* correlations assign each of the four *CH* units. Both techniques converge to establish the *CH* skeleton of the *ortho*-disubstituted benzene ring.

Reference to the deshielding of a ring proton by an *ortho* carboxy group clarifies the assignment.

^{13}C and ^1H chemical shifts δ_C and δ_H *(italic)*

2

CH multiplicities , CH coupling constants (Hz) , coupling protons

C-1	S		m						
C-2	S		d	7.5	(4-*H*)	d	7.5	(6-*H*)	("t")
C-3	D	165.4	d	6.7	(5-*H*)				
C-4	D	166.0	d	7.0	(6-*H*)				
C-5	D	162.4	d	7.3	(3-*H*)				
C-6	D	162.4	d	5.5	(4-*H*)				
C-7	S		b						
C-8	D	174.6	b						

HH multiplicities, HH coupling constants (Hz), coupling protons

3-H	d	7.5	(4-*H*)				
4-H	d	7.5	(3-*H*)	d	7.5	(5-*H*)	("t")
5-H	d	7.5	(4-*H*)	d	7.5	(6-*H*)	("t")
6-H	d	7.5	(5-*H*)				

33 Dicyclopentadiene

The ^{13}C NMR spectrum does not show the three resonances expected for monomeric cyclopentadiene. Instead, ten distinct signals appear, of which the DEPT spectrum identifies four *CH* carbon atoms in each of the shift ranges appropriate for alkanes and alkenes and in the alkane range additional two CH_2 carbon atoms. This fits the [4+2]-cycloadduct *2* of cyclopentadiene *1*.

1   *2*

The structure of the dimer can be derived simply by evaluation of the cross signals in the *HH* COSY plot. The cycloalkene protons form two *AB* systems with such small shift differences that the cross signals lie within the contours of the diagonal signals.

¹H chemical shifts δ_H assigned by *HH* COSY   ¹³C chemical shifts δ_C assigned by *HH* and *CH* COSY

The complete assignment of the C atoms follows from the *CH* correlation (*CH* COSY) and removes any uncertainty concerning the ¹³C signal assignments in the literature. The *endo*-linkage of the cyclopentene ring to the norbornene residue can be detected from the NOE on the protons at $\delta_H = 2.66$ and *3.12*, if the proton 7-H_{syn} at $\delta_H = 1.25$ is decoupled. Decoupling of the proton 7-H_{anti} at $\delta_H = 1.47$ leads to NOE enhancement of only the bridgehead protons at $\delta_H = 2.72$ and *2.80*.

34   *trans*-1-Cyclopropyl-2-methylbuta-1,3-diene

In the ¹³C NMR spectrum two signals with unusually small shift values [$(CH_2)_2$: $\delta_C = 7.7$; CH: $\delta_C = 10.6$] and remarkably large *CH* coupling constants ($J_{CH} = 161.9$ and 160.1 Hz) indicate a monosubstituted cyclopropane ring **A**. The protons which belong to this structural unit at $\delta_H = 0.41$ (*AA'*), *0.82* (*BB'*) and *1.60* (*M*) with typical values for *cis* couplings (*8.1 Hz*) and *trans* couplings (*4.9 Hz*) of the cyclopropane protons can be identified from the *CH* COSY plot.

¹³C and ¹H Chemical shifts δ_C and δ_H (*italic*)   *HH* coupling constants (Hz)

$^3J_{AM} = {}^3J_{A'M} = 4.9$ *(trans)*
$^3J_{BM} = {}^3J_{B'M} = 8.1$ *(cis)*
$^3J_{MX} = 9.8$

The additional coupling (*9.8 Hz*) of the cyclopropane proton X at $\delta_H = 1.60$ is the result of a *vicinal* H atom in the side-chain. This contains a methyl group **B**, a vinyl group **C** and an additional substituted ethenyl group **D**, as may be seen from the one dimensional ^1H and ^{13}C NMR spectra and from the CH COSY diagram.

Since the vinyl-CH proton at ($\delta_H = 6.33$ shows no additional $^3J_{HH}$ couplings apart from the doublet of doublets splitting (*cis* and *trans* coupling), the side-chain is a 1-isoprenyl chain **E** and not a 1-methylbuta-1,3-dienyl residue **F**.

Hence it must be either *trans*- or *cis*-1-cyclopropyl-2-methylbuta-1,3-diene (1-isoprenylcyclopropane), **G** or **H**.

In decoupling the methyl protons, the NOE difference spectrum shows a nuclear Overhauser enhancement on the cyclopropane proton at $\delta_H = 1.60$ and on the terminal vinyl proton with *trans* coupling at $\delta_H = 5.05$ and, because of the *geminal* coupling, a negative NOE on the other terminal proton at $\delta_H = 4.87$. This confirms the *trans* configuration **G**. In the *cis* isomer **H** no NOE would be expected for the cyclopropane proton, but one would be expected for the alkenyl-H in the α-position indicated by arrows in **H**.

^{13}C and ^1H chemical shifts δ_C and δ_H *(italic)*

CH multiplicities , CH coupling constants (Hz) , coupling *protons*

C-1′	D	160.1		m
C-2′,3′	T	161.9		m
C-1	D	150.4		m
C-2	S			m
C-3	D	151.3		d 8-0 (1-H) q 4-0 (CH₃)
C-4	D	158.7	D 153.9	
2-CH₃	Q	125.6		d 8.0 (1-H) d 4.4 (3-H)

HH multiplicities, HH coupling constants (Hz), coupling *protons*

1′-H$^{M'}$	d	9.8	(1-H)	t	8.1	(2′,3′-H$^{BB'}$)	t 4.9 (2′,3′-H$^{AA'}$)	
1-H	d	9.8	(1′-H)					
3-H	d	17.0	(4-H ^{E})	d	11.0	(4-H ^{Z})		
4-H (E)	d	17.0	(3-H)					
4-H (Z)	d	11.0	(3-H)					
CH₃	d	1.5	(1′-H)					

35   *cis*-6-Hydroxy-1-methyl-4-isopropylcyclohexene (carveol)

The correlation signals of the INADEQUATE experiment directly build up the ring skeleton **A** of the compound. Here characteristic ^{13}C shifts (δ_C = 123.1, 137.6; 148.9, 109.1) establish the existence and position of two double bonds and of one tetrahedral C–O single bond (δ_C = 70.5). DEPT spectra for the analysis of the *CH* multiplicities become unnecessary, because the INADEQUATE plot itself gives the number of CC bonds that radiate from each C atom.

The *CH* connectivities can be read off from the *CH* COSY plot; thus the complete pattern **B** of all *H* atoms of the molecule is established. At the same time an *OH* group can be identified by the fact that there is no correlation for the broad signal at δ_H = *4.45* in the *CH* COSY plot.

AB systems : a for axial, e for equatorial protons

The relative configuration of the *OH* and isopropenyl groups remains to be established. The ^1H signal at δ_H = *1.55*, a *CH$_2$* proton, splits into a pseudotriplet (*12.4 Hz*) of doublets (*10.1 Hz*). One of the two *12.4 Hz* couplings is the result of the other *geminal* proton of the *CH$_2$* group; the second of the two *12.4 Hz* couplings and the additional *10.1 Hz* coupling correspond to an *antiperiplanar* relationship of the coupling protons; the *vicinal* coupling partners of this methylene proton are thus located *axial* as depicted in the stereoformula **C**, with a *cis* configuration of the *OH* and isopropenyl groups. Hence it must be one of the enantiomers of carveol (**C**) shown in projection **D**.

36   *trans*-2-Methylcyclopentanol

Following the strategy applied in the previous problem, the correlation signals of the INADEQUATE experiment build up the methylcyclopentane skeleton **A** of the compound. DEPT subspectra **c** support the analysis of the *CH* multiplicities also resulting from the INADEQUATE plot which gives the number of CC bonds that radiate from each C atom.

Proton signals are assigned in **B** by means of the *CH* connectivities which can be read off from the *CH* COSY plot. Additionally, an *OH* group can be identified by the fact that there is no correlation for the broad signal at δ_H = *4.33* in the *CH* COSY plot. Connection of the OH group (δ_H =*4.33*) with the deshielded *CH* carbon atom (δ_C = 79.2) completes the structure of 2-methylcyclopentanol.

Signal overcrowding prevents an analysis of multiplets and coupling constants in the proton NMR spectrum so that the relative configuration is better derived from NOE difference spectra **e**. Decoupling of the methyl protons (δ_H = 0.74) leads to a significant NOE at the *vicinal CH* proton (δ_H = 3.43) and *vice versa*. The methyl protons and the *CH* proton are *cis* to each other; consequently, hydroxy function and methyl group adopt the *trans* configuration **C** (two enantiomers).

37   *trans*-2-(2-Pyridyl)methylcyclohexanol

The *CH* fragment which is linked to the *OH* group (δ_H = 5.45) can easily be located in the ${}^1H$ and ${}^{13}C$ NMR spectra. The chemical shift values δ_C =74.2 for C and δ_H = 3.16 for H are read from the *CH* COSY plot. The ${}^1H$ signal at δ_H = 3.16 splits into a triplet (*11.0 Hz*) of doublets (*4.0 Hz*). The fact that an *antiperiplanar* coupling of *11 Hz* appears twice indicates the *diequatorial configuration* (*trans*) of the two substituents on the cyclohexane ring **3**. If the substituents were positioned *equatorial-axial* as in **4** or **5**, then a *synclinal* coupling of ca *4 Hz* would be observed two or three times.

The pyridine chemical shifts can easily be assigned with the help of the *HH* coupling constants (cf. 3-acetylpyridine, Problem **6**). The ^{13}C chemical shift values of the bonded C atoms can then be read from the *CH* COSY plot. It is more difficult to assign the tetramethylene fragment of the cyclohexane ring because of signal overcrowding. The *geminal AB* systems of the individual CH_2 groups are clearly differentiated in the *CH* COSY plot; the *axial* protons (δ_H = *1.01-1.22*) show smaller ^1H shift values than their *equatorial* coupling partners on the same C atom as a result of anisotropic effects; they also show pseudoquartets because of two additional *diaxial* couplings. The *HH* COSY plot identifies the *HH* connectivities of the *H* atoms attached to the C-7–C-2 and C-1–C-6 subunits for structure **3**. Finally, the INADEQUATE plot differentiates between the CH_2 groups in positions 4 and 5 of the cyclohexane ring and confirms the aliphatic carbon skeleton.

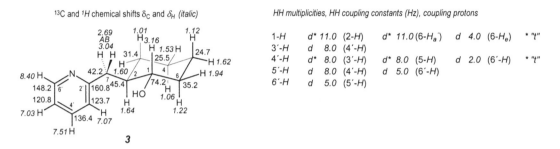

^{13}C and ^{1}H chemical shifts δ_C and δ_H (italic)

HH multiplicities, HH coupling constants (Hz), coupling protons

1-H	d* 11.0 (2-H)	d* 11.0 (6-H$_a$)	d 4.0 (6-H$_e$)	* "t"
3'-H	d 8.0 (4'-H)			
4'-H	d* 8.0 (3'-H)	d* 8.0 (5-H)	d 2.0 (6'-H)	* "t"
5'-H	d 8.0 (4'-H)	d 5.0 (6'-H)		
6'-H	d 5.0 (5'-H)			

3

7-$H^A H^B$ form an *AB* system ($^2J_{AB}$ = *14 Hz*) of doublets (H^A: ^3J = *5.0*; H^B: ^3J = *4.5 Hz*) as a result of coupling with 2-*H*.

38 Nona-2-*trans*-6-*cis*–dienal

From the *HH* COSY plot the following *HH* connectivities **A** are derived:

A   0.92 ···· 1.99 ···· 5.39 ···· 5.26 ···· 2.22 ···· 2.36 ···· 6.80 ···· 6.08 ···· 9.45 δ_H

In the *CH* COSY plot it can be established which C atoms are linked with these protons; thus the *CH* skeleton **B** can readily be derived from **A**:

B
0.92 ···· 1.99 ···· 5.39 ···· 5.26 ···· 2.22 ···· 2.36 ···· 6.80 ···· 6.08 ···· 9.45 δ_H
14.2 — 20.5 — 133.3 — 126.7 — 25.4 — 32.7 — 158.1 — 133.2 — 194.0 δ_C

Structural elucidation can be completed to give **C** if the *CH* multiplicities from the ^{13}C NMR spectrum and characteristic chemical shift values from the ^1H and ^{13}C NMR spectra are also taken into account. The *CH* bond with δ_C/δ_H = *194.0/9.45*, for example, clearly identifies an aldehyde group; *CH* bonds with δ_C/δ_H = *133.2/6.08, 158.1/6.80, 126.7/5.26* and *133.5/5.39* identify two CC double bonds of which one (δ_C/δ_H = *133.1/6.08* and *158.1/6.80*) is polarised by the –*M* effect of the aldehyde group.

C $\overset{9}{C}H_3 - \overset{8}{C}H_2 - \overset{7}{C}H = \overset{6}{C}H - \overset{5}{C}H_2 - \overset{4}{C}H_2 - \overset{3}{C}H = \overset{2}{C}H - \overset{1}{C}H = O$

Hence the compound is nona-2,6-dienal. The relative configuration of both CC double bonds follows from the *HH* coupling constants of the alkene protons in the ^1H NMR spectrum. The protons of the polarised 2,3-double bond are in *trans* positions ($^3J_{HH}$ = *15.5 Hz*) and those on the 6,7-double bond are in *cis* positions ($^3J_{HH}$ = *10.5 Hz*). The structure is therefore nona-2-*trans*-6-*cis*-dienal, **D**.

In assigning all shift values, *CH* and *HH* coupling constants, differentiation between C-2 and C-7 is at first difficult because the signals are too crowded in the ^{13}C NMR spectrum. Differentiation is possible, however, on closer examination of the *CH* COSY plot and the coupled ^{13}C NMR spectrum: the signal at δ_C = 133.2 splits as a result of *CH* long-range couplings into a doublet (25.0 Hz) of triplets (5.7 Hz), whose 'left' halves overlap in each case with the less clearly resolved long-range multiplets of the neighbouring signal, as the signal intensities show. Thereby, the coupling constant of 25.0 Hz locates the aldehyde proton which is two bonds apart from the C atom at δ_C = 133.2.

^{13}C and ^1H chemical shifts δ_C and δ_H *(italic)*

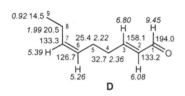

D

CH multiplicities , CH coupling constants (Hz) , coupling *protons*

C-1	D	171.0	d	9.5	(2-H)
C-2	D	160.2	d	25.0	(1-H)	t	5.7	(4-H$_2$)
C-3	D	151.3	t	5.5	(4-H$_2$)	t	5.5	(5-H$_2$)	("qui")
C-4	T	127.2	m
C-5	T	126.5	m
C-6	D	155.1	t	4.5	(4-H$_2$)	t	4.5	(5-H$_2$)	t	4.5	(8-H$_2$)	("sep")
C-7	D	158.3	o*
C-8	T	127.8	m
C-9	Q	126.5	m

HH multiplicities, HH coupling constants (Hz), coupling protons

1-H	d	7.9	(2-H)
2-H	d	15.5	(3-H, *trans*)	t	7.9	(1-H)		t	1.4	(4-H$_2$)
3-H	d	15.5	(2-H, *trans*)	t	6.9	(4-H$_2$)
4-H$_2$	d	6.9	(3-H)	t	6.9	(5-H$_2$)	("q")
5-H$_2$	d	7.0	(6-H)	t	7.0	(4-H$_2$)	("q")
6-H	d	10.5	(7-H, *cis*)	t	7.0	(5-H$_2$)		t	1.2	(8-H$_2$)
7-H	d	10.5	(6-H, *cis*)	t	7.0	(8-H$_2$)		t	1.4	(5-H$_2$)
8-H$_2$	d	7.0	(7-H)	q	7.0	(9-H$_3$)	("qui")	d	1.2	(6-H)
9-H$_3$	t	7.0	(8-H$_2$)

39   2,3-Diaza-7,8,12,13,17,18-hexaethylporphyrin

The ^1H NMR spectrum displays signals of shielded protons (δ_H = -2.35, integral level 1) and of deshielded ones (δ_H = *10.45* and *9.39*, integral levels 1 : 1). This reflects a ring current due to aromaticity as described for annulenes and porphyrins in section 2.5.2. To conclude, the reaction involves an oxidative cyclisation of 2,5-bis(2-pyrrolylmethyl)-1*H*-pyrrole *2* with 4*H*-triazole-3,5-dialdehyde *3* to the corresponding 2,3-diazaporphyrin *4*, following the "3+1" pathway of porphyrin synthesis. Two non-equivalent tautomers may exist; these are the diaza[18]annulene *4a* and the tetraaza[18]annulene *4b*.

Carbon atoms and protons are assigned by means of the proton-carbon connectivities as identified in the *H*C HSQC and HMBC experiment (**b** and **c**). The latter also permits the derivation of the connection of the ethyl groups to the porphyrin ring. The cross signals in the relevant part **a** of the *HH* COSY plot (**a**) are used to connect the methyl and methylene subunits to the ethyl groups.

3

2

$$-2H^+, -2e^-, -2H_2O$$

4a   **4b**

The equivalence of the inner N*H* protons ($\delta_H = -2.35$) as well as correlation signals with the pyrrolic carbons only ($\delta_C = 143.8, 141.6, 139.8$ and 136.7) provide evidence for the diaza[18]-annulene tautomer **4a**. Two separate N*H* proton signals and cross signals with the α-carbon atoms ($\delta_C = 159.3$) of the triazole ring are expected, in contrast, for the tetraaza[18]annulene tautomer **4b**.

proton-proton connectivities from *HH* COSY

3.97	1.88		3.88	1.82		3.73	1.76
—CH₂	—CH₃		—CH₂	—CH₃		—CH₂	—CH₃

proton-carbon connectivities from *HC* HSQC and HMBC

Proton	carbon atom attached	carbon atoms in two or three (or four) bonds distance			
δ_H	δ_C	δ_C			
10.45	102.2	136.7	143.8	159.3	
9.39	95.6	139.8	141.6	146.5	159.0
3.97	19.5	136.7	139.8	143.8	
3.88	19.3	139.8	141.6	143.8	
3.73	19.6	146.5	159.0		
1.88	18.1	143.8			
1.82	18.0	139.8			
1.76	18.2	146.5			
−2.35		143.8	141.6	139.8	136.7

¹³C and ¹H chemical shifts δ_C and δ_H (*italic*)

4a

40 2-Hydroxy-3,4,3',4'-tetramethoxydeoxybenzoin

First, nine double-bond equivalents from the molecular formula, twelve signals in the shift range appropriate for benzenoid C atoms and five multiplets in the shift range appropriate for benzenoid protons, with typical aromatic coupling constants, all indicate a double bond and two benzene rings. Of these two rings, one is 1,2,3,4-tetrasubstituted (*AB* system at $\delta_H = 6.68$ and 7.87 with an *ortho* coupling of *9 Hz*); the other is 1,2,4-trisubstituted (*ABC* system at $\delta_H = 6.79, 6.87$ and *6.97* with *ortho* and *meta* coupling, *8* and *2 Hz*, respectively). Substituents indicated include:

- in the ¹H NMR spectrum a phenolic O*H* group ($\delta_H = 12.34$),
- in the ¹³C NMR spectrum a ketonic carbonyl function ($\delta_C = 203.7$)
- and in both spectra four methoxy groups ($\delta_H = 3.68, 3.70, 3.71, 3.87$, and $\delta_C = 55.7, 56.3, 60.1$, respectively), in addition to a methylene unit ($\delta_H = 4.26$ and $\delta_C = 44.3$, respectively).

In order to derive the complete structure, the connectivities found in the *CH* COSY and *CH* CO-LOC plots are shown in Table 40.1.

Table 40.1. Proton-Carbon (J_{CH}) connectivities from the *CH* COSY and *CH* COLOC plots

Partial structure	Proton	carbon atom attached	carbon atoms in two or three (or four) bonds distance		
	δ_H	δ_C	δ_C		
A	7.87	128.0	203.7	158.6	156.5
B	6.68	104.0	136.0	114.5	
C	6.91	113.6	147.9	121.6	44.3
D	6.87	112.0	148.9	127.5	
E	6.79	121.6	147.9	113.6	
F	4.26	44.3	203.7	127.5	121.6 113.6
A	3.87	56.3	158.6		
D	3.71	55.7	148.9		
C	3.70	55.7	147.9		
B	3.68	60.1	136.0		

For the 1,2,3,4-tetrasubstituted benzene ring the partial structures **A** and **B** are derived from Table 40.1 from the connectivities of the *AB* protons at $\delta_H = 6.68$ and $\delta_H = 7.87$ and the methoxy protons at $\delta_H = 3.68$ and *3.87*. The complete arrangement of the C atoms of the second 1,2,4-trisubstituted benzene ring can be derived from the connectivities **C**, **D** and **E** of the protons of the *ABC* system ($\delta_H = 6.79, 6.87$ and *6.97*). From the partially resolved contours of the overlapping correlation signals at $\delta_C/\delta_H = 148.9/3.71$ and *147.9/3.70*, the methoxy protons at $\delta_H = 3.70$ and *3.71* can be attached to methoxy carbon atoms with the common ^{13}C signal at $\delta_C = 55.7$.

Finally, from the partial structures **A** and **F** it can be seen that the two benzene rings are linked to one another by a –CO–CH$_2$– unit ($\delta_C/\delta_H = 203.7$–$44.3/4.26$). Hence it must be 2-hydroxy-3,4-3',4'-tetramethoxydeoxybenzoin, **G**.

^{13}C and ^1H chemical shifts δ_C and δ_H (italic)

HH coupling constants (Hz), coupling protons
$^3J_{5,6} = 9$; $^3J_{5',6'} = 8$; $^4J_{2',6'} = 2$

41   3',4',7,8-Tetramethoxyisoflavone

The molecular formula contains ten double-bond equivalents. In the ^1H and ^{13}C NMR spectra four methoxy groups can be identified (δ_C = 61.2, 56.7, 57.8 and δ_H = 3.96, 3.87, 3.78, respectively). Of these, two have identical frequencies, as the signal intensity shows (δ_C = 55.8 and δ_H = 3.78). In the ^1H NMR spectrum an *AB* system (δ_H = 7.29 and 7.85) with *ortho* coupling (9 Hz) indicates a 1,2,3,4-tetrasubstituted benzene ring **A**; an additional *ABC* system (δ_H = 6.99, 7.12 and 7.19) with *ortho* and *meta* coupling (8.5 and 2 Hz) belongs to a second 1,2,4-trisubstituted benzene ring **B**. What is more, the ^{13}C NMR spectrum shows a conjugated carbonyl C atom (δ_C = 175.1) and a considerably deshielded *CH* fragment (δ_C = 154.0 and δ_H = 8.48) with the larger *CH* coupling (198.2 Hz) indicative of an enol ether bond, e.g. in a heterocycle such as furan, 4*H*-chromene or chromone **C**.

Knowing the substitution pattern of both benzene rings **A** and **B**, one can deduce the molecular structure from the *CH* connectivities of the *CH* COSY and *CH* COLOC plots. The interpretation of both experiments leads firstly to the correlation Table 41.1.

Table 41.1. Proton-Carbon (J_{CH}) connectivities from the *CH* COSY and *CH* COLOC plots

Partial structure	Proton δ_H	carbon atom attached δ_C	carbon atoms in two or three (or four) bonds distance δ_C		
C	8.48	154.0	175.1	150.1	123.3
A	7.85	121.2	175.1	156.4	150.1
A	7.29	111.2	136.3	118.7	
B	7.19	112.9	148.9	121.5	
B	7.12	121.5	148.9		
B	6.99	111.8	148.5	124.5	
A	3.96	56.7	156.4		
A	3.87	61.2	136.3		
B	3.78	55.8	148.9	148.5	

The benzene rings **A** and **B** derived from the ^1H NMR spectrum can be completed using Table 41.1. The way in which the enol ether is bonded is indicated by the correlation signal of the proton at δ_H = 8.48. The structural fragment **C** results, incorporating the C atom resonating at δ_C = 123.3, which has not been accommodated in ring **A** or **B** and which is two bonds ($^2J_{CH}$) removed from the enol ether proton.

The combination of the fragments **A-C** completes the structure and shows the compound to be 3',4',7,8-tetramethoxyisoflavone, **D**.

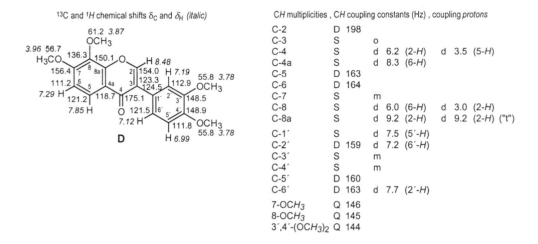

13C and 1H chemical shifts δ_C and δ_H (italic)

CH multiplicities , CH coupling constants (Hz) , coupling *protons*

C-2	D	198						
C-3	S		o					
C-4	S		d	6.2	(2-*H*)	d	3.5	(5-*H*)
C-4a	S		d	8.3	(6-*H*)			
C-5	D	163						
C-6	D	164						
C-7	S		m					
C-8	S		d	6.0	(6-*H*)	d	3.0	(2-*H*)
C-8a	S		d	9.2	(2-*H*)	d	9.2	(2-*H*) ("t")
C-1'	S		d	7.5	(5'-*H*)			
C-2'	D	159	d	7.2	(6'-*H*)			
C-3'	S		m					
C-4'	S		m					
C-5'	D	160						
C-6'	D	163	d	7.7	(2'-*H*)			
7-OCH_3	Q	146						
8-OCH_3	Q	145						
3',4'-(OCH_3)_2	Q	144						

42   3',4',6,7-Tetramethoxy-3-phenylcoumarin

Isoflavones *3* that are unsubstituted in the 2-position are characterised in their ^1H and ^{13}C NMR spectra by two features:

- a carbonyl-C atom at $\delta_C \sim 175$ (cf. problem 41);
- an enol ether C*H* fragment with high ^1H and ^{13}C chemical shift values ($\delta_H/\delta_C \sim 8.5/154$) and a remarkably large $^1J_{CH}$ coupling constant (~ 198 Hz, cf. problem 41).

The NMR spectra of the product do not show these features. The highest ^{13}C shift value is $\delta_C = 160.9$ and indicates a conjugated carboxy-C atom instead of the keto carbonyl function of an isoflavone ($\delta_C = 175$). On the other hand, a deshielded C*H* fragment at $\delta_C/\delta_H = 138.7/7.62$ appears in the ^{13}C NMR spectrum, which belongs to a CC double bond polarised by a $-M$ effect. The two together point to a coumarin *4* with the substitution pattern defined by the reagents.

The correlation signals of the C*H* COSY and the C*H* COLOC plots (shown in the same diagram) confirm the coumarin structure *4*. The carbon and hydrogen chemical shifts and couplings indicated in Table 42.1 characterise rings **A**, **B** and **C**. The connection of the methoxy protons also follows easily from this experiment. The assignment of the methoxy C atoms remains unclear because their correlation signals overlap. Hence the correspondence between the methoxy double signal at $\delta_C = 55.8$ and the 3',4'-methoxy signals ($\delta_C = 55.8$) of 3',4',6,7-tetramethoxyisoflavone (problem 41) may be useful until experimental proof of an alternative is found.

Table 42.1. Proton-carbon (J_{CH}) connectivities from the CH COSY and CH COLOC plots

Partial structure	Proton δ_H	carbon atom attached δ_C	carbon atoms in two or three (or four) bonds distance δ_C			
C	7.62	138.7	160.9	148.9	127.6	107.8
B	7.22	111.5	149.2	124.3	120.8	
B	7.16	120.8	149.2	124.3	120.8	
A	6.83	107.8	152.3	148.9	146.2	138.7
B	6.81	110.8	148.5	127.6		
A	6.72	99.3	152.3	148.9	146.2	112.2
A	3.86	56.2	152.3			
B	3.85	55.8	148.5			
A	3.84	56.2	146.2			
B	3.82	55.8	149.2			

A   C   B

^{13}C and ^1H chemical shifts δ_C and δ_H (italic)

D

CH multiplicities , CH coupling constants (Hz) , coupling protons

C-2	s	d 8.0 (4-H)			
C-3	s	d 4.0 (2'-H)	d 4.0 (6'-H)		("t")
C-4	D 160	d 6.0 (5-H)			
C-4a	s	d 6.0 (8-H)			
C-5	D 160	d 4.0 (4-H)			
C-6	s	d 7.5 (8-H)	d 3.7 (5-H)	q 3.7 (OCH$_3$)	(d"qui")
C-7	s	d 8.0 (5-H)	d 4.0 (8-H)	q 4.0 (OCH$_3$)	(d"qui")
C-8	D 162				
C-8a	s	o			
C-1'	s	d 8.0 (5'-H)	d 4.0 (4-H)		
C-2'	D 158	d 8.0 (6'-H)			
C-3'	s	d 8.0 (5'-H)	d 4.0 (2'-H)	q 4.0 (OCH$_3$)	(d"qui")
C-4'	s	o			
C-5'	D 160				
C-6'	D 163	d 8.0 (2'-H)	d 1.0 (5'-H)		
6,7-(OCH$_3$)$_2$	Q 145				
3',4'-(OCH$_3$)$_2$	Q 145				

43   Aflatoxin B$_1$

The keto-carbonyl ^{13}C signal at $\delta_C = 200.9$ would only fit the aflatoxins B$_1$ and M$_1$. In the ^{13}C NMR spectrum an enol ether-CH fragment can also be recognised from the chemical shift value of $\delta_C = 145.8$ and the typical one-bond coupling constant $J_{CH} = 196$ Hz; the proton involved appears at $\delta_H = 6.72$, as the CH COSY plot shows. The ^1H triplet which belongs to it overlaps with a singlet, identified by the considerable increase in intensity of the central component. The coupling constant of the triplet *2.5 Hz* is repeated at $\delta_H = 5.39$ and *4.24*. Judging from the *CH* COSY plot, the proton at $\delta_H = 5.39$ is linked to the C atom at $\delta_C = 102.5$ (Table 43.1); likewise, on the basis of its shift value it belongs to the β-C atom of an enol ether fragment, shielded by the *+M* effect of the enol ether O atom. The other coupling partner, the allylic proton at $\delta_H = 4.24$, is linked to the C atom at $\delta_C = 47.1$, as can be seen from the *CH* COSY plot (Table 43.1). It appears as a doublet (*7 Hz*) of pseudotriplets (*2.5 Hz*). The larger coupling constant (*7 Hz*) reoccurs in the doublet at $\delta_H = 6.92$. According to the *CH* COSY plot (Table 43.1) the C atom at $\delta_C = 113.5$ is bonded to this proton. Hence the evidence tends towards partial structure **A**, and so away from aflatoxin M_1, in which the allylic proton would be substituted by an *OH* group.

coupling protons and HH coupling constants (Hz, **bold**)

δ_H	6.72	5.39	4.24	6.92
6.72	—	**2.5**	**2.5**	
5.39	**2.5**	—	**2.5**	
4.24	**2.5**	**2.5**	—	**7.0**
6.92		**7.0**		—

Table 43.1. Proton-carbon (J_{CH}) connectivities from the *CH* COSY and *CH* COLOC plots

Partial structure	Proton δ_H	carbon atom attached δ_C	carbon atoms in two or three (or four) bonds distance δ_C			
C	2.46	34.9	200.9			
C	3.22	28.8	177.4			
B	3.91	57.2	161.4			
A	4.24	47.1				
A	5.39	102.5	145.8			
A	6.72	145.8	113.5	102.5	47.1	
B	6.72	91.4	165.1	161.4	107.2	103.5
A	6.92	113.5				

Further interpretation of the *CH* COSY / *CH* COLOC plots allows additional assignments to be made for fragments **B** and **C** of aflatoxin B_1.

Since fragment **A** was clearly assigned with the help of *HH* coupling constants, all of the C atoms not included in **A**, which, according to the *CH* COLOC plot, are two or three bonds apart from the equivalent protons at $\delta_H = 6.72$ (Table 43.1), belong to the benzene ring **B**.

The assignment of the quaternary C atoms at $\delta_C = 154.3$, 152.1 and 116.4 has yet to be established. The signal with the smallest shift ($\delta_C = 116.4$) is assigned to C-11a because the substituent effects of carboxy groups on α-C atoms are small. Since the signal at $\delta_C = 152.1$ in the coupled spectrum displays a splitting ($^3J_{CH}$ coupling to 9a-H), it is assigned to C-3c.

Additional evidence for the assignment of the other C atoms is supplied by the *CH* coupling constants in the Table shown.

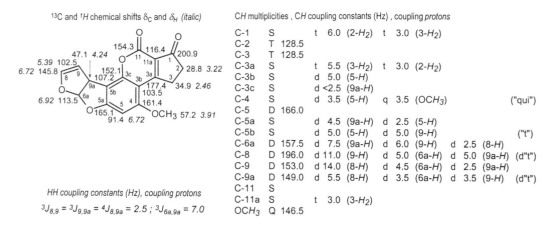

^{13}C and ^{1}H chemical shifts δ_C and δ_H *(italic)*

HH coupling constants (Hz), coupling protons

$^{3}J_{8,9} = {}^{3}J_{9,9a} = {}^{4}J_{8,9a} = 2.5$; $^{3}J_{6a,9a} = 7.0$

CH multiplicities , CH coupling constants (Hz) , coupling *protons*

C-1	S		t	6.0	(2-*H*₂)	t	3.0	(3-*H*₂)		
C-2	T	128.5								
C-3	T	128.5								
C-3a	S		t	5.5	(3-*H*₂)	t	3.0	(2-*H*₂)		
C-3b	S		d	5.0	(5-*H*)					
C-3c	S		d	<2.5	(9a-*H*)					
C-4	S		d	3.5	(5-*H*)	q	3.5	(OC*H*₃)		("qui")
C-5	D	166.0								
C-5a	S		d	4.5	(9a-*H*)	d	2.5	(5-*H*)		
C-5b	S		d	5.0	(5-*H*)	d	5.0	(9-*H*)		("t")
C-6a	D	157.5	d	7.5	(9a-*H*)	d	6.0	(9-*H*)	d 2.5 (8-*H*)	
C-8	D	196.0	d	11.0	(9-*H*)	d	5.0	(6a-*H*)	d 5.0 (9a-*H*)	(d"t")
C-9	D	153.0	d	14.0	(8-*H*)	d	4.5	(6a-*H*)	d 2.5 (9a-*H*)	
C-9a	D	149.0	d	5.5	(8-*H*)	d	3.5	(6a-*H*)	d 3.5 (9-*H*)	(d"t")
C-11	S									
C-11a	S		t	3.0	(3-*H*₂)					
OC*H*₃	Q	146.5								

44   1,5-Dimethylcyclohexa-1,3-dien-5-ol-6-one, dimer

The ^{13}C NMR spectrum of the metabolite shows 16 signals instead of 8 as expected from the elemental composition determined by high-resolution mass spectrometry. Moreover, aromaticity of the 2,6-xylenol is obviously lost after metabolism because two ketonic carbonyl carbon atoms (δ_C = 203.1 and 214.4) and four instead of twelve carbon signals are observed in the shift range of trigonal carbon nuclei (δ_C = 133.1, 135.4, 135.6 and 139.4) in the ^{13}C NMR spectra. To conclude, metabolism involves oxidation of the benzenoid ring.

The connectivities found in the *HH* COSY diagram reveal two partial stuctures **A** and **B** according to Table 44.1. These are supported by the identical proton-proton coupling constants of the coupling protons.

Table 44.1. Proton-proton (J_{HH}) connectivities and partial structures **A** and **B** from the *HH* COSY plots

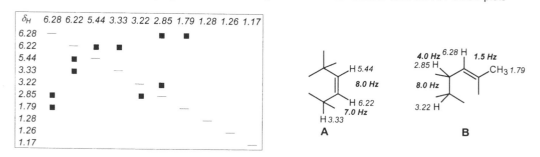

Carbon-proton bonds are then assigned by means of the one-bond *CH* connectivities observed in the *CH* COSY diagram (Table 44.2). This completes partial structures **A** and **B** to the *CH* skeletons **C** and **D**.

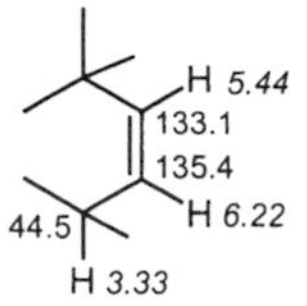

Structures C and D

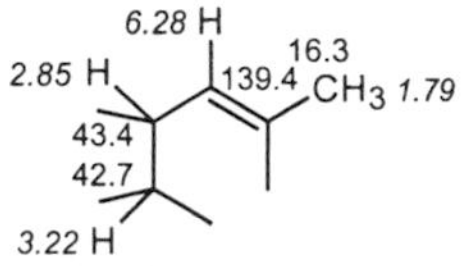

Table 44.2. Proton-carbon (J_{CH}) connectivities from the CH COSY and CH COLOC plots

CH$_n$	Proton	carbon atom attached	carbon atoms in two or three bonds distance		
	δ_H	δ_C	δ_C		
CH	6.28	139.4			
CH	6.22	135.4			
CH	5.44	133.1			
CH	3.33	44.5			
CH	3.22	42.7			
CH	2.85	43.4			
CH$_3$	1.79	16.3	203.1	139.4	135.6
CH$_3$	1.28	15.7	214.4	133.1	53.7
CH$_3$	1.26	31.7	203.1	73.7	42.7
CH$_3$	1.17	26.1	214.4	72.6	

Two- and three-bond CH connectivities of the methyl protons detected in the CH COLOC diagram identify the partial structures **E** and **F**. Only one bond from $\delta_C = 72.6$ to $\delta_C = 44.5$ could not be directly identified because of the absence of the corresponding cross signal ($\delta_H = 1.17$ with $\delta_C = 44.5$). Nevertheless, partial structure **E** is reasonable, assuming the six-membered ring to be retained during metabolism.

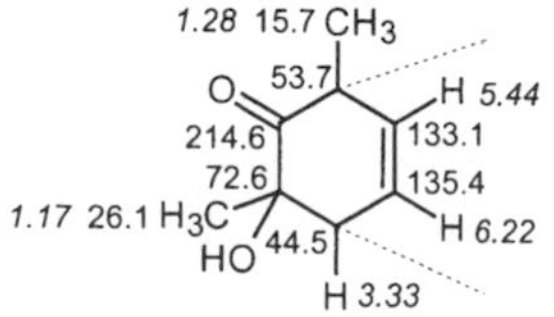

Structures E and F

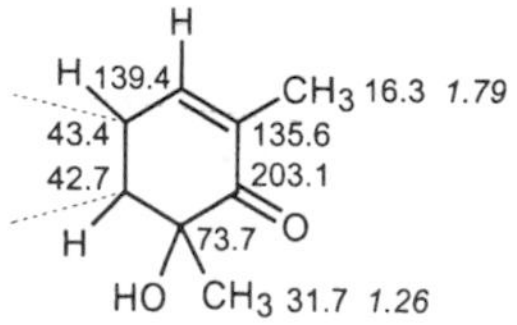

Fragments **E** and **F** which include all 16 carbon atoms detected by ^{13}C NMR can be attached to each other in two ways **G** or **H**; the structure **G** is realised, as follows from the NOE difference spectra, which show a significant NOE between the methyl protons at $\delta_H = 1.28$ and the alkene proton at $\delta_H = 6.28$ and vice versa (Table 44.3).

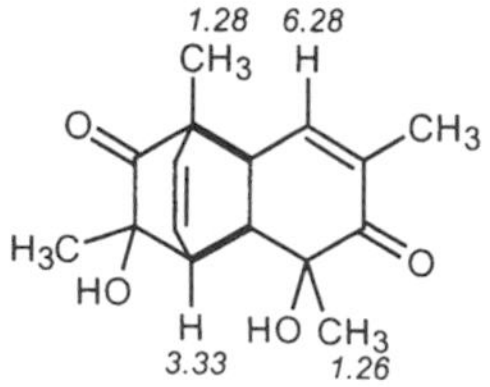

Structures G and H

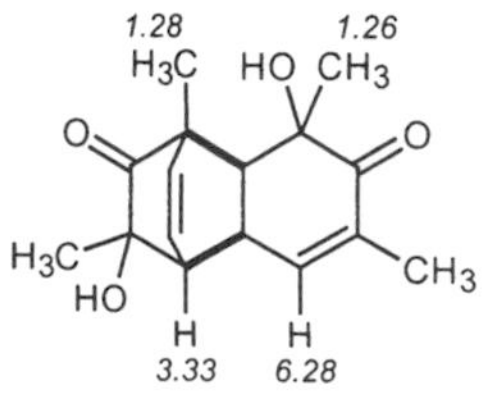

Table 44.3. Significant proton-proton NOEs from the *HH* NOE difference spectra

δ_H	6.28	6.22	5.44	3.33	3.22	2.85	1.79	1.28	1.26	1.17
6.28	—					+	+	+		
6.22		—	+	+						+
5.44		+	—						+	
3.33		+		—	+					+
3.22				+	—	+			+	
2.85	+				+	—		+		
1.79	+						—			
1.28	+		+			+		—		
1.26					+				—	
1.17		+		+						—

An NOE between the the alkene proton at $\delta_H = 6.22$ and the methyl protons at $\delta_H = 1.17$ establishes the relative configuration (*exo*) of the respective methyl group. The *exo* attachment of the six-membered ring in the stereostructure **I** follows, in particular, from the NOE between the methyl protons at $\delta_H = 1.26$ and the bridgehead proton at $\delta_H = 3.22$ as well as the absence of effects between the alkenyl proton pair with $\delta_H = 5.44/6.22$ and the bridgehead proton pair with $\delta_H = 2.85/3.22$.

Also the mirror image of the structure **I**, correctly denoted as *exo*-3,10-dihydroxy-3,5,8,10-tetra-methyltricyclo[6.2.2.0^{2,7}]dodeca-5,11-diene-4,9-dione, would be possible since enantiomers are not differentiated by NMR. A retro-Diels-Alder fragmentation of **I** to $C_8H_{10}O_4$ explains why the molecular ion corresponding to the molecular formula $C_{16}H_{20}O_4$ is not detected in the mass spectrum. The metabolite **I** could be formed by Diels-Alder dimerisation of 1,5-dimethylcyclohexa-1,3-dien-5-ol-6-one **J** as the primary metabolite which acts as diene and dienophile as well [39].

45   Asperuloside

The molecular formula $C_{18}H_{22}O_{11}$ contains eight double-bond equivalents, i.e. four more than those in the framework *1* known to be present. The ^{13}C NMR spectrum shows two carboxy-CO double bonds (δ_C = 170.2 and 169.8) and, apart from the enol ether fragment (C-3: δ_C = 148.9, J_{CH} = 194.9 Hz; C-4: δ_C = 104.8, +M effect of the ring O atom), a further CC double bond (C: δ_C = 142.9; CH: δ_C = 127.3); the remaining double-bond equivalent therefore belongs to an additional ring.

Analysis of the CH correlation signals (CH COSY / CH COLOC) for the protons at δ_H = 7.38 and 5.54 (Table 45.1) shows this ring to be a five-membered lactone. The CH correlation signals with the protons at δ_H = 4.65 (AB system of methylene protons on C-10) and δ_H = 2.04 (methyl group) identify and locate an acetate residue (CO: δ_C = 170.2; CH_3: δ_C = 20.8) at C-10 (Table 45.1).

Table 45.1. Partial structures from the CH COSY and CH COLOC plots (the protons are given in *italic* numerals, C atoms separated by a single bond are given in bold numerals and C atoms separated by two or three bonds are given in small ordinary numerals

CH correlation maxima with the hydrogen atoms at δ_H = 5.70, 5.54, 4.65, 3.55 and 3.22 finally establish the position of the additional CC double bond (C-7/C-8, Table 45.1). Hence the basic structure **A** of the aglycone is now clear.

The iridoid monoterpenoid part of the structure C-1–C-9–C-5–C-6–C-7 (**B**) is confirmed by the *HH* COSY plot:

The ^1H and ^{13}C signal assignments of glucopyranoside ring **C** are derived from the *HH* COSY and *CH* COSY diagram:

As can be seen from a Dreiding model, the five- and six-membered rings of **A** only link *cis* so that a bowl-shaped rigid fused-ring system results. Protons 5-*H*, 6-*H* and 9-*H* are in *cis* positions and therefore almost eclipsed. The relative configuration at C-1 and C-9 has yet to be established. Since 1-*H* shows only a very small $^3J_{HH}$ coupling (*1.5 Hz*) which is scarcely resolved for the coupling partner 9-*H* (δ_H = 3.22), the protons are located in such a way that their *CH* bonds enclose a dihedral angle of about 120°. The O-glucosyl bond is therefore positioned *synclinal* with respect to 9-*H*.

The *antiperiplanar* coupling constant (*8 Hz*) of the protons 1'-*H* (δ_H = *4.49*) and 2'-*H* (δ_H = *2.98*) finally shows that a β-glucoside is involved.

The assignment of all of the chemical shift values and coupling constants as derived from the measurements can be checked in structural formula **D**.

The natural product is the asperuloside described in the literature [40]. The assignments for the carbon pairs C-1 / C-1' , C-6' / C-1 and C-11 / CO (acetyl) have been interchanged. Deviations of ^{13}C chemical shifts (CDCl$_3$–D$_2$O [40]) from the values tabulated here [(CD$_3$)$_2$SO] are due mainly to solvent effects. Here the difference between the measurements **a** and **d** shows that the use of D$_2$O exchange to locate the O*H* protons where the *CH* COSY plot is available is unnecessary since O*H* signals give no *CH* correlation signal. In this case D$_2$O exchange helps to simplify the *CH–OH* multiplets and so interpretation of the *HH* COSY plot, which only allows clear assignments when recorded at 600 MHz.

CH multiplicities , *CH* coupling constants (Hz) , coupling *protons*

C-1	D	179.8	b							
C-3	D	194.9	b							
C-4	S		d	2.0	(5-*H*)	d	2.0	(6-*H*)	d	2.0 (9-*H*) ("q")
C-5	D	149.3	b							
C-6	D	164.9								
C-7	D	169.8	b							
C-8	S		b							
C-9	D	137.6	d	9.5	(7-*H*, trans)					
C-10	T	148.0								
C-11	S		d	3.0	(3-*H*, cis)					
C-1'	D	160.3	b							
C-2'	D	138.9	b							
C-3'	D	137.6	b							
C-4'	D	144.2	b							
C-5'	D	140.7	b							
C-6'	T	140.5	b							
Ac-CO	S		q	5.0	(CH$_3$)					
Ac-CH$_3$	Q	129.7								

HH coupling constants (Hz), where resolved

$^3J_{1,9}$ = 1.5 ; $^4J_{3,5}$ = 2.0 ; $^3J_{5,6}$ = 8.0 ; $^3J_{5,9}$ = 8.0 ; $^2J_{10\text{-}AB}$ = 14.0 ; $^3J_{1',2'}$ = 8.0 (anti) ; $^3J_{2',3'}$ = 7.5 (anti) ;
$^3J_{3',4'}$ = 7.5 (anti) ; $^3J_{4',5'}$ = 8.0 (anti) ; $^3J_{5',6'}$ = 8.0 (anti) ; $^3J_{5',6'}$ = 3.0 (syn) ; $^2J_{6'\text{-}AB}$ = 12.5

46   Lacto-*N*-tetrose

The $^3J_{HH}$ coupling constants (*8.2, 8.0,* and *8.4 Hz*) of the anomeric protons (1-*H*, δ_H = *5.14, 4.57,* and *4.44*) reveal a *diaxial configuration* of the protons attached to C-1 and C-2 in all pyranose units, reflecting three β-glycosidic linkages. The connectivities found in the *HH* COSY diagram **a** permit to distinguish between the four peracetylated sugar residues galactosyl^1, galactosyl^2, 2-*N*-acetylamino-2-deoxyglucopyranosyl- (GlcNac), and 2,3,6-tri-*O*-acetyl-sorbityl- (reduced gluco-pyranosyl):

β-D-galactopyranosyl- (Gal ^{1})

β-D-2-*N*-acetylamino-2-deoxy-D-glucopyranosyl-

β-D-galactopyranosyl- (Gal ^{2})

β-D-2,3,6-tri-*O*-acetyl-D-sorbityl-

These assignments are confirmed by the *HH* TOCSY diagram **b** and the selective one-dimensional experiment on top of **b** with the ^1H NMR subspectrum of the 2-*N*-acetylamino-2-deoxygluco-pyranosyl- residue.

The sequence of the sugar units is deduced from the *HH* ROESY in comparison with the *HH* TOCSY plot. This is conveniently achieved by copying the ROESY experiment on transparent paper, putting both experiments one above the other and by looking for additional cross signals of the anomeric protons in the *HH* ROESY. Additional cross signals (spatial correlations) of the anomeric protons are detected between δ_H = *4.44 / 4.58* (Gal^1-1-*H* / GlcNac-3-*H*), *4.57 / 4.10* (Gal^2-1-*H* / Sorb-4-*H*), and *5.14 / 3.78* (GlcNac-1-*H* / Gal^2-3-*H*).

To conclude, the sequence of sugars is Gal^1-β1–3-GlcNac-β1–3-Gal^2-β1–4-Sorbitol (peracetyla-ted):

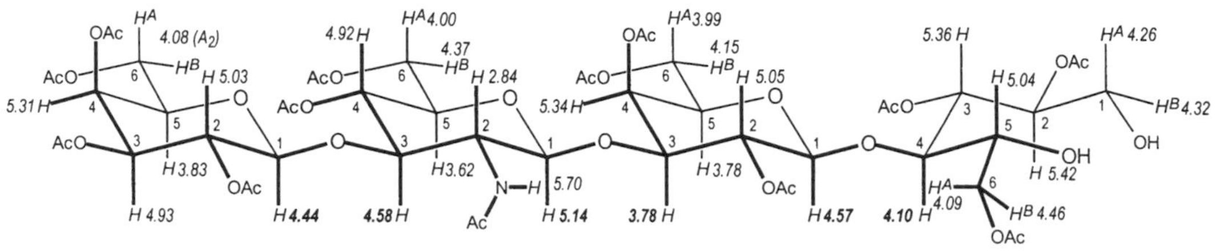

Additionally, all carbon-proton bonds can be assigned by means of the *HC* HSQC experiment **b**:

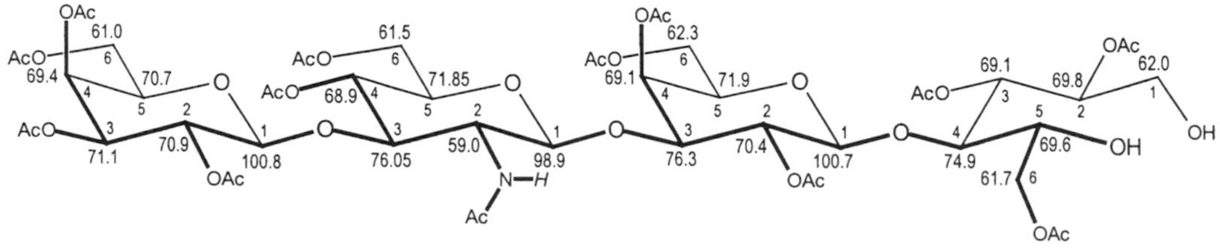

Finally, the structure of the original (genuine) tetrasaccharide (before acetylation and reduction) is Gal^1-β1–3-GlcNac-β1–3-Gal^2-β1–4-Glc, known as Lacto-*N*-tetrose:

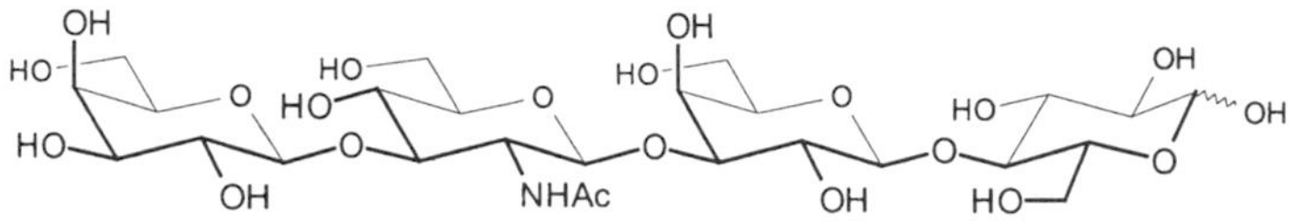

47   9-Hydroxycostic acid

In the ^1H broadband decoupled ^{13}C NMR spectrum, 15 carbon signals can be identified, in agree-ment with the molecular formula which indicates a sesquiterpene. The DEPT experiments show that the compound contains four quaternary C atoms, three *CH* units, seven *CH$_2$* units and a *CH$_3$*

group (Table 47.1); this affords the CH partial formula $C_{15}H_{20}$. Consequently, two H atoms are not linked to carbon. Since the molecular formula contains oxygen as the only heteroatom, these two H atoms belong to OH groups (alcohol, carboxylic acid). The ^{13}C NMR spectrum shows a carboxy C atom ($\delta_C = 170.4$). In the solvent (CD_3OD) the carboxylic proton is not observed because of deuterium exchange. According to CH COSY and DEPT, the second OH group belongs to a secondary alcohol ($CHOH$) with the shifts $\delta_C = 80.0$ and $\delta_H = 3.42$ (Table 47.1).

In the alkene shift range, two methylene groups are found, whose CH connectivities are read off from the CH COSY plot (Table 47.1, $=CH_2$: δ_C/δ_H = 123.4 / *5.53 AB 6.18* and $=CH_2$: δ_C/δ_H = 106.9 / *4.47 AB 4.65*). The quaternary alkene C atoms to which they are bonded appear in the ^{13}C NMR spectrum at $\delta_C = 146.9$ and 151.1 (Table 47.1). Because of the significant difference in the chemical shift values, one of the two CC double bonds ($\delta_C = 123.4$) must be more strongly polarised than the other ($\delta_C = 106.9$), which suggests that it is linked to the carboxy group ($-M$ effect). The carboxy function and the two $C=CH_2$ double bonds together give three double-bond equivalents. In all, however, the molecular formula contains five double-bond equivalents; the additional two evidently correspond to two separate or fused rings.

Table 47.1. Intepretation of the CH COSY plot (CH bonds)

δ_c	CH_n	attached proton(s) δ_H
170.4	COO	
151.0	C	
146.9	C	
123.4	CH_2	*5.53 AB 6.18*
106.9	CH_2	*4.47 AB 4.65*
80.0	CH	*3.42*
49.8	CH	*1.88*
42.3	C	
38.9	CH_2	*1.23 AB 1.97*
38.5	CH	*2.60*
37.8	CH_2	*2.05 AB 2.32*
36.5	CH_2	*1.53 AB 1.79*
30.8	CH_2	*1.33 AB 1.60*
24.5	CH_2	*1.55 AB 1.68*
11.2	CH_3	*0.75*
CH partial formula	$C_{15}H_{20}$	

Two structural fragments **A** and **B** can be deduced from the HH COSY plot; they include the AB systems of *geminal* protons identified from the CH COSY diagram (Table 47.1). Fragments **A** and **B** can be completed with the help of the CH data in Table 47.1.

A δ_H *2.05(A) / 2.32(B)*   *1.55(A) / 1.68(B)*   *1.23(A) / 1.97(B)*
       δ_C 37.8 —— 24.5 —— 38.9

B δ_H *3.42*   *1.53(A) / 1.79(B)*   *2.60*   *1.33(A) / 1.60(B)*   *1.88*
       δ_C 80.0 —— 36.5 —— 38.5 —— 30.8 —— 38.9 ——

The way in which **A** and **B** are linked can be deduced from the CH COLOC plot. There it is found that the C atoms at $\delta_C = 80.0$ (CH), 49.8 (CH), 42.3 (C) and 38.9 (CH_2) are separated by two or three bonds from the methyl protons at ($\delta_H = 0.75$) and thus structural fragment **C** can be derived.

In a similar way, the linking of the carboxy function with a CC double bond follows from the correlation of the carboxy resonance (δ_C = 170.4) with the alkene protons (δ_H = 5.53 and 6.18); the latter give correlation signals with the C atom at δ_C = 38.5, as do the protons at δ_H = 1.33 and 1.53, so that taking into account the molecular unit **B** which is already known, an additional substructure **D** is established.

The position of the second CC double bond in the structural fragment **E** follows finally from the correlation of the ^{13}C signals at δ_C = 37.8 and 49.8 with the ^1H signals at δ_H = 4.47 and 4.65. Note that *trans* protons generate larger cross-sectional areas than *cis* protons as a result of larger scalar couplings.

Table 47.2. Assembly of the partial structures **A**-**E** to form the decalin framework **F** of the sesquiterpene

Table 47.2 combines partial structures **A, B, C, D** and **E** into the decalin framework **F**. The relative configurations of the protons can be derived from an analysis of all the *HH* coupling constants in the expanded ^1H multiplets. The *trans*-decalin link is deduced from the *antiperiplanar* coupling (*12.5 Hz*) of the protons at δ_H = 1.33 and 1.88. The *equatorial* configuration of the OH group is derived from the doublet splitting of the proton at δ_H = 3.42 with *12.5* (*anti*) and *4.5 Hz* (*syn*). In a corresponding manner, the proton at δ_H = 2.60 shows a pseudotriplet (*12.5 Hz*, two *anti* protons) of pseudotriplets (*4.0 Hz*, two *syn* protons), whereby the *equatorial* configuration of the 1-carboxyethenyl group is established. Assignment of all *HH* couplings, which can be checked in Table 47.3 provides the relative configuration **G** of all of the ring protons in the *trans*-decalin. The stereoformula **G** is the result; its mirror image would also be consistent with the NMR data. Formula **G** shows the stronger shielding of the *axial* protons compared with their *equatorial* coupling partners on the same C atom and combines the assignments of all ^{13}C and ^1H shifts given in Table 47.1. The result is the known compound 9β-hydroxycostic acid [41].

Table 47.3. Relative configurations of the protons between $\delta_H = 1.23$ and 3.42 from the *HH* coupling constants of the expanded proton multiplets. Chemical shift values (δ_H) of the proton multiplets are given as large numerals in boldface and coupling constants (Hz) are as small numerals

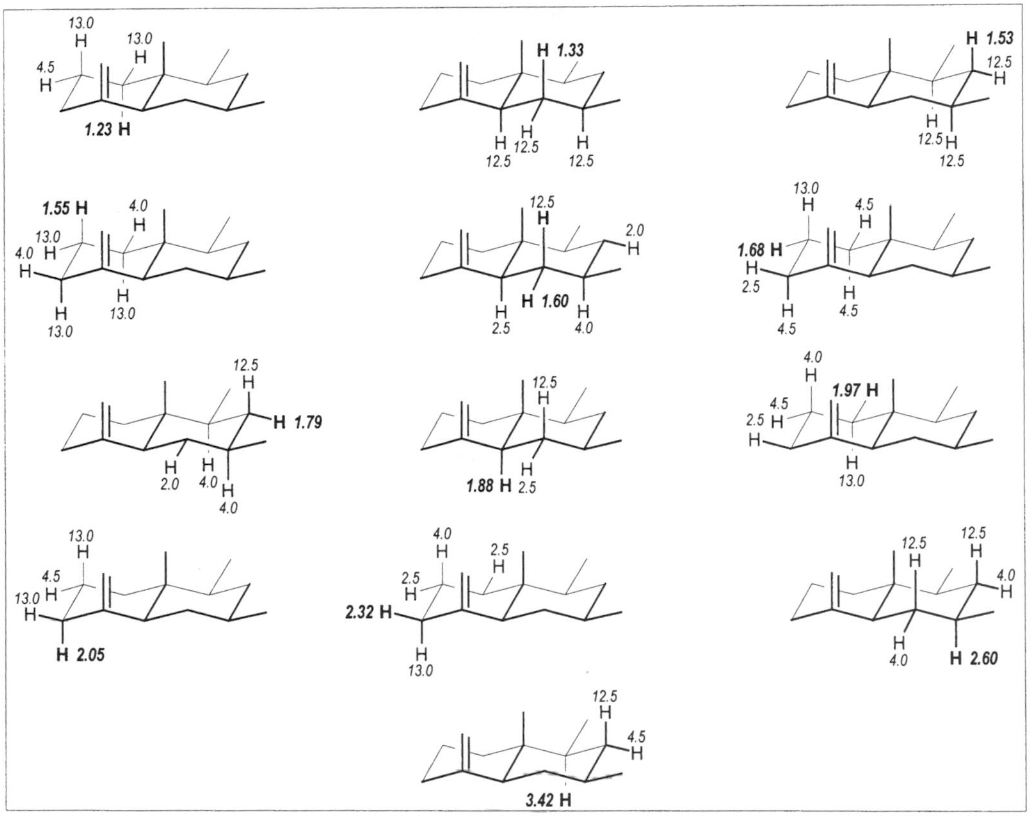

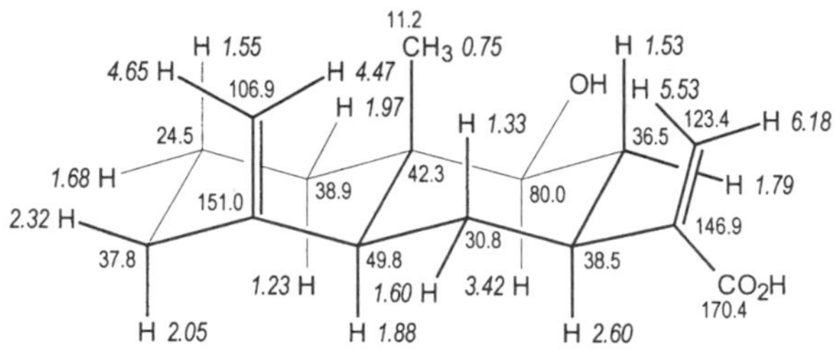

48   14-(Umbelliferon-7-*O*-yl)driman-3,8-diol

The given structure **A** is confirmed by interpretation of the C*H* COSY and C*H* COLOC diagrams. All of the essential bonds of the decalin structure are derived from the correlation signals of the methyl protons. In this, the DEPT subspectra differentiate between the tetrahedral C atoms which are bonded to oxygen (δ_C = 75.5: CH–O; 72.5: C–O; 66.6: CH$_2$–O). The methyl protons at δ_H = 1.19, for example, give correlation maxima with the C atoms at δ_C = 72.5 ($^2J_{CH}$), 59.4 ($^3J_{CH}$) and 44.1 ($^3J_{CH}$). A corresponding interpretation of the other methyl CH correlations ($^3J_{CH}$ relationships) gives the connectivities which are indicated in bold in structure **A**. The assignment of CH$_2$ groups in positions 2 and 6 remains to be established; this can be done by taking into account the deshielding (α- and β-effects and the shielding γ-effects (as sketched in formulae **B** and **C**).

The assignment of the umbelliferone residue in **A** likewise follows from interpretation of the J_{CH} and $^{2,3}J_{CH}$ relationships in the CH COSY and CH COLOC plots following Table 48.1. The ^{13}C signals at δ_C = 112.9 and 113.1 can be distinguished with the help of the coupled ^{13}C NMR spectrum: δ_C = 112.9 (C-3′) shows no $^3J_{CH}$ coupling, whereas δ_C = 113.1 (C-6′) shows a $^3J_{CH}$ coupling of 6 Hz to the proton 8′-H.

Table 48.1. Interpretation of the CH COSY and CH COLOC plots

| Protons | | C atoms separated by | | | | |
| | | one bond | | two or three bonds | | |
δ_H	CH$_n$[a]	δ_C	δ_C	δ_C	δ_C	δ_C
7.59	CH	143.5	161.3	155.7	128.7	
7.30	CH	128.7	161.8	155.7	143.5	
6.82	CH	101.6	161.8	155.7		
6.80	CH	113.1	161.8	155.7		
6.19	CH	112.9	161.3			
4.13 AB 4.37	CH$_2$	66.6				
3.39	CH	75.5				
1.53 AB 1.90	CH$_2$	44.1[b]	32.7			
1.53 AB 1.90	CH$_2$	25.1[b]				
1.84	CH	59.4				
1.39 AB 1.65	CH$_2$	32.7				
1.30 AB 1.55	CH$_2$	20.0				
1.49	CH	48.8				
1.19	CH$_3$	24.6	72.5	59.4	44.1	
0.96	CH$_3$	28.4	75.5	48.4	37.4	22.1
0.90	CH$_3$	16.0	59.4	48.4	37.9	32.7
0.80	CH$_3$	22.1	75.5	48.4	37.9	28.4

[a] CH multiplicities from the DEPT ^{13}C NMR spectra.
[b] AB systems of the protons attached to these C atoms overlap.

Because of signal overcrowding in the aliphatic range between δ_H = 1.3 and 2.0, the HH coupling constants cannot be analysed accurately. Only the deshielded 3-H at δ_H = 3.39 shows a clearly recognisable triplet fine structure. The coupling constant of *2.9 Hz* indicates a dihedral angle of 60° with the protons 2-H^A and 2-H^B ; thus, 3-H is *equatorial*. If it were *axial* then a double doublet with one larger coupling constant (ca *10 Hz* for a dihedral angle of 180°) and one smaller coupling constant (*3 Hz*) would be observed.

The NOE difference spectra provide more detailed information regarding the relative configuration of the decalin. First, the *trans* decalin link can be recognised from the significant NOE of the methyl ^1H signals at $\delta_H = 0.80$ and *0.90*, which reveals their *coaxial* relationship as depicted in **D**. For *cis* bonding of the cyclohexane rings an NOE between the methyl protons at $\delta_H = 0.90$ and the *cis* bridgehead proton 5-II ($\delta_H = 1.49$) would be observed, as **E** shows for comparison. An NOE between the methyl protons at $\delta_H = 0.90$ and *1.19* proves their *coaxial* relationship, so the 8-O*H* group is *equatorial*.

Further effects confirm what has already been established (5-H at $\delta_H = 1.49$ *cis* to the methyl protons at *0.96*; 3-H at $\delta_H = 3.39$ *cis* to the methyl group at $\delta_H = 0.80$; 14-$H^A H^B$ with H^A at $\delta_H = 4.13$ and H^B at *4.37* in spatial proximity to the umbelliferone protons 6'-H and 8'-H at $\delta_H = 6.80$ and *6.82*). The natural product is therefore 14-(umbelliferon-7-O-yl)-driman-3α,8α-diol, **D**, or its enantiomer.

Stereoformulae **F** (with ^1H chemical shifts) and stereoprojection **G** (with ^{13}C chemical shifts) summarise all assignments, whereby *equatorial* protons exhibit the larger ^1H shifts according to their doublet structure which can be detected in the C*H* COSY plot; *equatorial* protons, in contrast to their coupling partners on the same C atom, show only *geminal* couplings, and no additional comparable *antiperiplanar* couplings. The NOE difference spectra also differentiate between the O–C*H*$_2$ protons ($\delta_H = 4.13$ close to the coaxial methyl group at *0.90*, $\delta_H = 4.37$ close to the methyl group at *1.19* as shown in **F**).

HH coupling constants (Hz), where resolved

$^3J_{2A,3} = {}^3J_{2B,3} = 2.9$; $^2J_{14A,14B} = 9.9$; $^3J_{9,14A} = 5.6$; $^3J_{9,14B} = 4.0$;
$^3J_{3',4'} = 9.5$; $^3J_{5',6'} = 8.6$; $^4J_{6',8'} = 2.5$

CH multiplicities , CH coupling constants (Hz) , coupling *protons*

C-1	T	126.2							
C-2	T	125.7							
C-3	D	146.2							
C-4	S								
C-5	D	122.0							
C-6	T	124.7							
C-7	T	125.0							
C-8	S								
C-9	D	124.7							
C-10	S								
C-11	Q	125.2							
C-12	Q	125.7							
C-13	Q	125.0							
C-14	T	143.6							
C-15	Q	123.1							
C-2′	S								
C-3′	D	173.1							
C-4′	D	163.1	d	5.2	(5′-H)				
C-4a′	S		m						
C-5′	D	162.0	d	3.7	(4′-H)				
C-6′	D	163.1	d	5.2	(8′-H)				
C-7′	S		d	10.0	(5′-H)				
C-8′	D	163.6	d	4.7	(6′-H)				
C-8a′	S		d	10.0	(5′-H)	d	5.8	(4′-H)	d 5.8 (8′-H) ("t")

G (δ$_C$)

49   3,4,5-Trimethyl-5,6-dihydronaphtho[2,3-*b*]furan

The molecular formula $C_{15}H_{16}O$, which indicates a sesquiterpene, contains eight double bond equivalents; in the sp^2 carbon chemical shift range (δ$_C$ = 107.5 - 154.4) ten signals appear which fit these equivalents. Since no carboxy or carbonyl signals can be found, the compound contains five CC double bonds. Three additional double bond equivalents then show the system to be tricyclic.

In the ^{13}C NMR spectrum the large CH coupling constant (197.0 Hz) of the CH signal at δ$_C$ = 141.7 indicates an enol ether unit (=CH–O–), as occurs in pyran or furan rings. The long-range quartet splitting ($^3J_{CH}$ = 5.9 Hz) of this signal locates a CH$_3$ group in the α-position of this carbon atom. This structural element **A** occurs in furanosesquiterpenes, the furanoeremophilanes.

A

Starting from the five CC double bonds, three rings and a 3-methylfuran structural fragment, analysis of the CH COSY and CH COLOC diagrams leads to Table 49.1 and the identification of fragments **B-J**.

CH coupling relationships over two and three bonds (very rarely more) cannot always be readily identified. However, progress can be made with the help of the CH fragments which have been identified from the CH COSY plot, and by comparing the structural fragments **B - J** with one another. One example would be the assignment of the quaternary C atoms at δ$_C$ = 130.0 and 133.2 in the fragments **G** and **H**: the weak correlation signals with the proton at δ$_H$ = 6.54 may originate from CH couplings over two or three bonds; the correlation signal δ$_C$/δ$_H$ = 130.0/5.94 clarifies this; the alkene proton obviously only shows CH relationships over three bonds, namely to the C atoms at δ$_C$ = 130.0 and 27.5.

Table 49.1. Interpretation of the *CH* COSY and *CH* COLOC plots

Partial structure	Protons		C atoms separated by						
			one bond		two or three bonds				
	δ_H	$CH_n{}^a$	δ_C	δ_C	δ_C	δ_C	δ_C	δ_C	δ_C
B	1.16	CH_3	19.6	27.5	31.1	133.2			
C	2.30 AB 2.63	CH_2	31.1	19.6	27.5	125.3	128.2	133.2	
D	2.44	CH_3	11.4	116.5	126.6	141.7			
E	2.63	CH_3	14.1	107.5	125.3	126.6	127.9	130.0	133.2
F	3.36	CH	27.5	19.6	125.3	130.0	133.2		
G	5.94	CH	125.3	27.5	130.0				
H	6.54	CH	128.2	31.1	107.5	130.0	133.2		
I	7.05	CH	107.5	126.6	128.2	133.2	154.4		
J	7.33	CH	141.7	116.5	126.6	154.4			

ᵃ *CH* multiplicities from the DEPT ^{13}C NMR spectra.

The structural fragments **B - J** converge to 3,4,5-trimethyl-5,6-dihydronaphtho[2,3-*b*]furan, **K**. Whether this is the 5(*S*)-or 5(*R*)-enantiomer (as shown) cannot be decided conclusively from the NMR measurements. It is clear, however, that the 5-C*H* proton at $\delta_H = 3.36$ is split into a pseudo-quintet with *7.1 Hz*; this is only possible if one of the 6-CH_2 protons (at $\delta_H = 2.63$) forms a dihedral angle of about 90° with the 5-C*H* proton so that $^3J_{HH} \approx 0$ *Hz*.

HH coupling constants (Hz) where resolved
$^4J_{2,3-Me} = 1.5$; $^3J_{5,6A} = 7.1$; $^3J_{5,5-Me} = 7.1$;
$^2J_{6A,6B} = 16.5$; $^3J_{6A,7} = 1.5$; $^3J_{6B,7} = 6.5$;
$^4J_{6B,8} = 3.1$; $^3J_{7,8} = 9.5$

CH multiplicities , *CH* coupling constants (Hz) , coupling *protons*

C-2	D	197.0	q	5.9 (3-CH_3)					
C-3	S		d	12.0 (2-H)	q	5.9 (3-CH_3)			
C-3a	S		m						
C-4	S		m						
C-4a	S		m						
C-5	D	130.0	m						
C-6	T	128.5	m						
C-7	D	161.5	d*	7.9 (5-H)	t*	7.9 (6-H_2)			*("q")
C-8	D	157.5	d*	5.9 (9-H)	t*	5.9 (6-H_2)			*("q")
C-8a	S		m						
C-9	D	161.5	d	3.7 (8-H)					
C-9a	S		d	7.9 (9-H)					
3-CH_3	Q	128.0							
4-CH_3	Q	126.0							
5-CH_3	Q	126.0	d	5.9 (6-H^A)	d*	3.0 (5-H)	d* 3.0 (6-H^B)	*("t")	

50   Sendarwine

The high-resolution molecular mass of the compound gives the molecular formula $C_{21}H_{28}O_6$, which corresponds to eight double bond equivalents. The ^1H broadband decoupled ^{13}C NMR spectrum shows a keto carbonyl group ($\delta_C = 185.2$), two carboxy functions ($\delta_C = 176.4$ and 170.4) and four further signals in the sp^2 chemical shift range ($\delta_C = 146.8$, 145.2, 134.3 and 120.9). These signals identify five double bonds. The three double bond equivalents still missing must then belong to three separate or fused rings. A complete interpretation of the *CH* COSY and DEPT experiments leads to the correlation Table 50.1 and to a *CH* partial molecular formula of $C_{21}H_{28}$, which shows that all 28 *H* atoms of the molecule are bonded to C atoms, and that no *OH* groups are present.

Table 50.1. Interpretation of the *CH* COSY plot and the DEPT spectra

carbon atoms δ_C	multiplicity CH_n [a]	attached protons δ_H
185.2	CO [b]	
176.4	COO [b]	
170.4	COO [b]	
146.8	=C	
145.2	=CH	7.31
134.3	=C	
120.9	=C	
75.8	CH–O	6.29
75.0	CH–O	4.88
54.9	CH	2.41
49.5	C	
44.1	CH	1.95
34.5	CH	2.62
29.5	CH_2	1.49 AB 1.95
21.2	CH_3	2.02
18.7	CH_3	1.21
18.6	CH_3	1.21
15.7	CH_2	1.66 AB 2.06
14.6	CH_3	0.92
9.8	CH_3	1.08
8.8	CH_3	1.83
	$C_{21}H_{28}$ [c]	

[a] *CH* multiplicities are obtained from the DEPT ^{13}C NMR subspectra.
[b] Other linkages are eliminated on the basis of the molecular formula.
[c] The C_xH_y partial formula is obtained by adding all CH_n units.

In the *HH* COSY plot, structural fragment **A** can be identified starting from the signal at $\delta_H = 4.88$. It is evident that two non-equivalent protons overlap at $\delta_H = 1.95$. The *CH* COSY diagram (expanded section) shows that one of these protons is associated with the *CH* at $\delta_C = 44.1$ and the other with the methylene C atom at $\delta_C = 29.5$. Altogether the molecule contains two CH_2 groups, identified in the DEPT subspectrum, whose methylene *AB* protons can be clearly analysed in the *CH* COSY plot and which feature as *AB* systems in the structural fragment **A**.

Resonances in the sp^2 carbon shift region provide further useful information: one at δ_C = 185.5 indicates a keto carbonyl function in conjugation with a CC double bond; two others at δ_C = 176.4 and 170.4 belong to carboxylic acid ester groups judging by the molecular formula and since O*H* groups are not present; four additional signals in the sp^2 shift range (δ_C = 146.8, 145.2, 134.3 and 120.9) indicate two further CC double bonds, hence the ^1H shift of δ_H = *7.31* and the C*H* coupling constant (202.2 Hz) of the ^{13}C signals at δ_C = 145.2 identify an enol ether fragment, e.g. of a furan ring with a hydrogen atom attached to the 2-position.

At this stage of the interpretation, the C*H* correlations across two or three bonds (C*H* COLOC plot) provide more detailed information. The ^1H shifts given in the C*H* COLOC diagram, showing correlation maxima with the C atoms at a distance of two to three bonds from a particular proton, lead to the recognition of eight additional structural fragments **B-I** (Table 50.2).

Table 50.2. Partial structures from the C*H* COLOC plot. Each partial structure **B-I** is deduced from the two- or three-bond couplings J_{CH} for the *H* atoms of **B-I** (with *italic* δ_H values)

Whether the C and *H* atoms as coupling partners are two or three bonds from one another (^2J or ^3J coupling) is decided by looking at the overall pattern of the correlation signals of a particular C atom with various protons. Thus for methyl protons at δ_H = *1.08*, correlation maxima for C atoms are found at δ_C = 54.9 (C*H*) and 49.5 (quaternary C). The proton which is linked to the C atom at δ_C = 54.9 (δ_H = *2.41*, cf. C*H* COSY diagram and Table 50.1) shows a correlation signal with the methyl C at δ_C = 9.8, which for its part is linked to the methyl protons at δ_H = *1.08*. From this the fragment **G**, which features parts of **C**, follows directly. The combination of all fragments (following Table 50.2) then gives the furanoeremophilane skeleton **J** of sendarwine.

The relative configuration of the protons follows from the $^3J_{HH}$ coupling constants, of which it is necessary to concentrate on only two signals (at δ_H = *4.88* and *2.41*). The proton at δ_H = *4.88* shows a quartet with a small coupling constant (*3 Hz*) which thus has no *antiperiplanar* relationship to one of the *vicinal* protons; it is therefore *equatorial* and this establishes the *axial* position of the acetoxy group. The C*H* proton at $\delta_H = 2.41$ shows an *antiperiplanar* coupling (*9 Hz*) and a *synclinal* coupling (*3 Hz*) with the neighbouring methylene protons. From this the relative configuration **K** or its mirror image is derived for the cyclohexane ring.

At first the configuration of the methyl groups at C-4a and C-5 remains unclear. The NOE difference spectra, which arise from the decoupling of various *axial* protons, provide the answer. Irradiation at $\delta_H = 1.49$ leads to NOE enhancement of the *coaxial* protons (*1.95* and *2.41*) and of the *cis* protons ($\delta_H = 4.88$). Irradiation at $\delta_H = 1.66$ has a strong effect on the methyl group at $\delta_H = 1.08$, and from this the *coaxial* relationship of these protons in the sense of three-dimensional structure **L** is the result. Decoupling at $\delta_H = 6.29$ induces strong effects on the *coaxial* protons at $\delta_H = 1.95$ and *2.41* and weak effects on the obviously distant methyl groups ($\delta_H = 0.92$ and *1.08*); irradiation at $\delta_H = 2.41$ has a corresponding effect, producing a very distinct NOE at $\delta_H = 6.29$ and a weaker effect at $\delta_H = 1.49$ and *1.95*, because in these signals the effects are distributed among multiplet lines. From the *coaxial* relationships thus indicated the structure **L** (or its mirror image with *cis* methyl groups in positions 4a and 5) is deduced.

The stereo projection **M** showing all ^1H and ^{13}C signals summarises all assignments. Again it is evident that *axial* protons (*a*) on the cyclohexane ring are more strongly shielded than their *equatorial* coupling partners (*e*) on the same C atom and that the diastereotopism of the isobutyric acid methyl groups is only resolved in the ^{13}C NMR spectrum.

Thus, sendarwine [42] is systematically named as 6β-acetoxy-4,4a,5,6,7,8,8a,9-octahydro-3,4aβ,5β-trimethyl-9-oxonaphtho[2,3-*b*]furan-4β-yl-2-methylpropanoic acid ester.

51   Panaxatriol

The sample prepared is not particularly pure, so instead of the 30 signals expected, 33 signals are observed in the ^1H broadband decoupled ^{13}C NMR spectrum. Only by pooling information from the DEPT experiment and from the CH COSY plot can a reliable analysis be obtained, as shown in Table 51.1. Here the AB systems of the *geminal* CH_2 protons are assigned.

The three H atoms present in the molecular formula $C_{30}H_{52}O_4$ but missing from the CH balance $C_{30}H_{49}$ (Table 51.1) belong to three hydroxy groups.

Table 51.1. Interpretation of the DEPT spectrum (CH_n) and the CH COSY plot

carbon atoms δ_C	multiplicity CH_n [a]	attached protons δ_H	carbon atoms δ_C	multiplicity CH_n [a]	attached protons δ_H
78.5	CH	3.15	36.5	CH$_2$	1.34 AB 1.50
76.6	C		35.7	CH$_2$	1.22 AB 1.55
73.2	C		33.1	CH$_3$	1.20
69.8	CH	3.50	31.1	CH$_2$	1.03 AB 1.45
68.6	CH	4.08	30.9	CH$_3$	1.30
61.1	CH	0.87	30.3	CH$_2$	1.18 AB 1.90
54.7	CH	1.90	27.2	CH$_3$	1.25
51.1	C		27.1	CH$_2$	1.55 AB 1.64
49.4	CH	1.40	25.2	CH$_2$	1.18 AB 1.78
48.7	CH	1.60	19.4	CH$_3$	1.16
47.0	CH$_2$	1.53 AB 1.55	17.2	CH$_3$	0.92
41.0	C		17.2	CH$_3$	1.04
39.3	C		17.1	CH$_3$	0.88
39.2	C		16.3	CH$_2$	1.55 AB 1.77
38.7	CH$_2$	1.01 AB 1.71	15.5	CH$_3$	0.92
			CH partial formula $C_{21}H_{28}$		

[a] CH multiplicities are obtained from the DEPT ^{13}C NMR subspectra.

Further information is derived from the NOE difference spectra with decoupling of the methyl protons. Table 51.2 summarises the most significant NOE enhancements to complete the picture.

Table 51.2. Interpretation of the NOE difference spectra

Irradiation δ_H	significant nuclear Overhauser enhancement (+) 0.87	0.92	0.97	1.04	1.30	1.40	1.60	3.15	3.50	4.08
0.88					+				+	
0.92		+	+							+
0.97				+						+
1.04		+					+			+
1.16							+			
1.30	+		+					+		

NOE enhancements (Table 51.2) reflect *coaxial* relationships between

- the CH–O proton at $\delta_H = 4.08$ and the CH_3 protons at $\delta_H = 0.92$, 0.97 and 1.04,
- the methyl group at $\delta_H = 0.88$ and the CH protons at $\delta_H = 1.40$ and 3.50,
- the CH proton at $\delta_H = 1.60$ and the CH_3 protons at $\delta_H = 1.04$ and 1.16,

as well as the *cis* relationship of the *CH* protons at $\delta_H = 0.87$ and *3.15* with respect to the methyl group at *1.30*. From this the panaxatriol structure **A** is derived starting from the basic skeleton of dammarane.

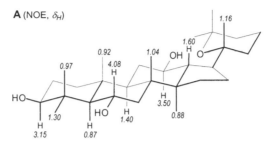

A (NOE, δ_H)

The multiplets and coupling constants of the (*axial*) protons at $\delta_H = 3.15$, *3.50* and *4.08* moreover confirm the *equatorial* positions of all three *OH* groups, as can be seen in formula **B**. Here the couplings from *10.0* to *11.5 Hz*, respectively, identify *vicinal* protons in *diaxial* configurations, whilst values of *4.5* and *5.0 Hz*, respectively, are typical for *axial–equatorial* relationships. As the multiplets show, the protons at $\delta_H = 3.50$ and *4.08* couple with two *axial* and one *equatorial* proton (triplet of doublets) respectively, whereas the proton at $\delta_H = 3.15$ couples with one *axial* and one *equatorial* proton (doublet of doublets).

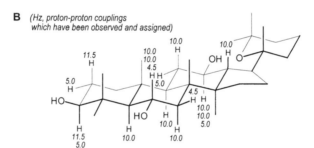

B   *(Hz, proton-proton couplings which have been observed and assigned)*

Well separated cross signals of the *HH* COSY plot demonstrate

- the *geminal* positions of the methyl groups at $\delta_H = 0.97$ and *1.30* and
- the *vicinal* relationship of the protons at $\delta_H = 3.15-$ *(1.55 AB 1.64)*, *1.60–3.50–(1.18 AB 1.90)* and *0.87–4.08–(1.53 AB 1.55)*.

Those C atoms which are bonded to the protons that have already been located can be read from the *CH* COSY plot (Table 51.1) and thus partial structure **C** is the result.

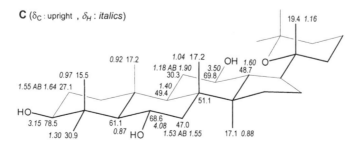

C (δ_C : upright , δ_H : *italics*)

The *CH* COLOC diagram shows correlation signals for the methyl protons which are particularly clear (Table 51.3). Interpretation of these completes the assignments shown in formula **D** by reference to those *CH* multiplicities which have already been established (Table 51.1).

Table 51.3. Interpretation of the *CH* COLOC diagram (methyl connectivities) using the *CH* multiplets derived from Table 51.1

Methyl protons δ_H	C atoms separated by two or three bonds							
	δ_C	CH_n	δ_C	CH_n	δ_C	CH_n	δ_C	CH_n
0.88	31.1	CH_2	41.0	C	48.7	CH	51.1	C
0.92	38.7	CH_2	39.2	C	49.4	CH	61.1	CH
0.97	30.9	CH_3	39.3	C	61.1	CH	78.5	CH
1.04	41.0	C	47.0	CH_2	49.4	CH	51.1	C
1.16	35.7	CH_2	54.7	CH	76.6	C		
1.20	27.2	CH_3	36.5	CH_2	73.2	C		
1.25	33.1	CH_3	36.5	CH_2	73.2	C		
1.30	15.5	CH_3	39.3	C	61.1	CH	78.5	CH

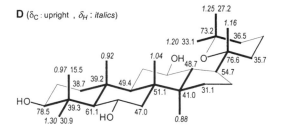

D (δ_C : upright , δ_H : italics)

Table 51.3 and formula **D** show that the methyl connectivities of the *CH* COLOC plot are sufficient to indicate essential parts of the triterpene structure.

Differentiation between the methyl groups at $\delta_C = 27.2$ and 33.1 and between the methylene ring C atoms at $\delta_C = 16.3$ and 25.2 remains. Here the γ effect on the ^{13}C chemical shift proves its value as a criterion: C-23 is more strongly shielded ($\delta_C = 16.3$) by the two *axial* methyl groups in (γ) positions 20 and 25 of the tetrahydropyran rings than is C-16 ($\delta_C = 25.2$). The *axial CH$_3$* group at C-25 is correspondingly more strongly shielded ($\delta_C = 27.2$) than the *equatorial* (33.1), in accordance with the reverse behaviour of the methyl protons. Thus formula **E** is derived with its complete assignment of all protons and carbon-13 nuclei.

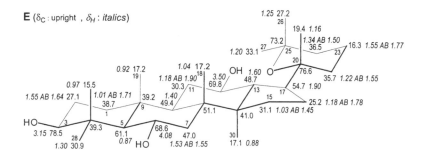

E (δ_C : upright , δ_H : italics)

52 *trans-N*-Methyl-4-methoxyproline

Rather large *HH* coupling constants in the aliphatics range (*12.5* and *15.0 Hz*) indicate *geminal* methyl protons in rings. In order to establish clearly the relevant *AB* systems, it makes sense first to interpret the *CH* COSY diagram (Table 52.1). From this, the compound contains two methylene groups, **A** and **B**.

A δ_H 2.34 AB 2.66   **B** δ_H 3.45 AB 4.02
      δ_C 38.2 δ_C 75.4

Table 52.1. Intepretation of the *CH* COSY and the *CH* COLOC plots

| Partial structure | *Protons* | | C atoms separated by | | |
	δ_H	$CH_n{}^a$	one bond δ_C	two or three bonds δ_C	δ_C
	4.58	CH	67.7		
	4.32	CH	77.8	170.6	
B	3.45 AB 4.02	CH_2	75.4	67.7	
	3.40	OCH_3	55.0	67.7	
	3.15	NCH_3	49.1	77.8	75.4
A	2.34 AB 2.66	CH_2	38.2	77.8	

ª *CH* multiplicities from the coupled and DEPT ^{13}C NMR spectra.

Taking these methylene groups into account, interpretation of the *HH* COSY plot leads directly to the *HH* relationships **C** even if the protons at $\delta_H = 2.34$ and *4.58* do not show the expected cross signals because their intensity is spread over the many multiplet lines of these signals.

                              A   **B**
                             2.34^A 3.45^A
C δ_H 4.32 → 2.66^B → 4.58 → 4.02^B

The *CH* COSY plot completes the *HH* relationships **C** of the *CH* fragment **D**:

                             2.34^A 3.45^A
D δ_H 4.32   2.66^B 4.58   4.02^B
        δ_C ──77.8── 38.2 ── 67.7 ── 75.4 ──
                     ── CH ── CH_2 ── CH ── CH_2 ──

The typical chemical shift values and *CH* coupling constants in the one-dimensional NMR spectra reveal three functional groups:

- *N*-methylamino (–NCH_3, $\delta_C = 49.1$; Q, $J_{CH} = 144$ Hz; $\delta_H = = 3.15$),
- methoxy (–OCH_3, $\delta_C = 55.0$; Q, $J_{CH} = 146$ Hz; $\delta_H = 3.40$),
- carboxy-/carboxamido- (–COO, $\delta_C = 170.6$).

If it were a carboxylic acid, the carboxy proton would not be visible because of deuterium exchange in the solvent tetradeuteriomethanol:

$$-CO_2H + CD_3OD \rightleftharpoons -CO_2D + CD_3OH$$

In the *CH* COLOC plot (Table 52.1) the correlation signals of the N–CH_3 protons ($\delta_H = 3.15$) with the terminal C atoms of the fragment **D** ($\delta_C = 75.4$ and 77.8) indicate an *N*-methylpyrrolidine ring **E**.

E

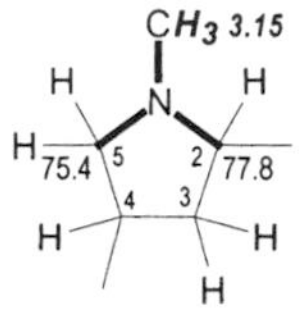

Since the carboxy-C atom in the *CH* COLOC diagram (Table 52.1) shows no correlation signal with the methoxy protons, it must be a carboxylic acid rather than a methyl ester. In the *CH* CO-LOC plot of problem 53 there is, for example, a cross signal relating the carboxy-C atom with the O*CH₃* protons, because this compound is a methyl ester. Finally, a cross signal relating the carb-oxy-C atom (δ_C = 170.6) to the ring proton 2-*H* (δ_H = *4.32*) in the *CH* COLOC plot locates the carboxy group on C-2. Hence the skeletal structure has been established; it is *N*-methyl-4-me-thoxyproline, **F**.

F

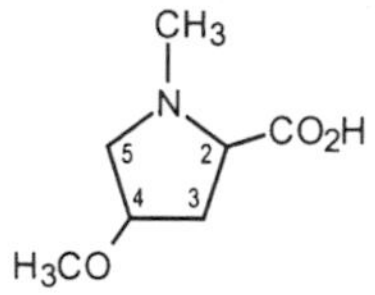

The relative configuration is derived from the NOE difference spectra. Significant NOEs are de-tected owing to *cis* relationships within the neighbourhood of non-*geminal* protons:

$$\delta_H = 2.34 \ / \ 4.32; \ \ 2.66 \ / \ 4.58; \ \ 4.02 \ / \ 4.58; \ \ (NCH_3) \ 3.15 \ / \ 4.02$$

From this, the *N*-methyl and carboxy groups are in *cis* positions whereas the carboxy and methoxy groups are *trans* and so *trans*-*N*-methyl-4-methoxyproline, **G**, is the structure implied. The NMR measurements do not provide an answer as to which enantiomer it is, 2*R*,4*S* as shown or the mirror image 2*S*,4*R*.

G

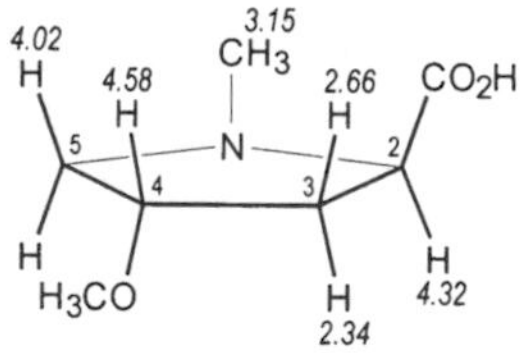

The formulae **H** and **I** summarize the results with the complete assignments of all ¹³C and ¹H che-mical shifts (**H**) and the *HH* multiplets and coupling constants (**I**). Here the ¹H multiplets which have been interpreted because of their clear fine structure are indicated by the multiplet abbrevia-tion *d* for *doublet*.

Chemical shift assignments (δ_C : upright , δ_H : *italics*)

H

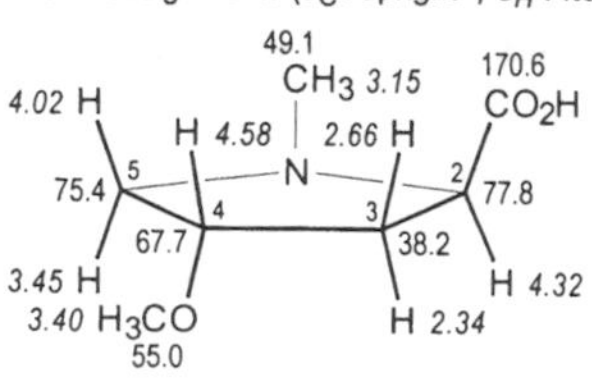

HH multiplicities and coupling constants (Hz) where resolved

I

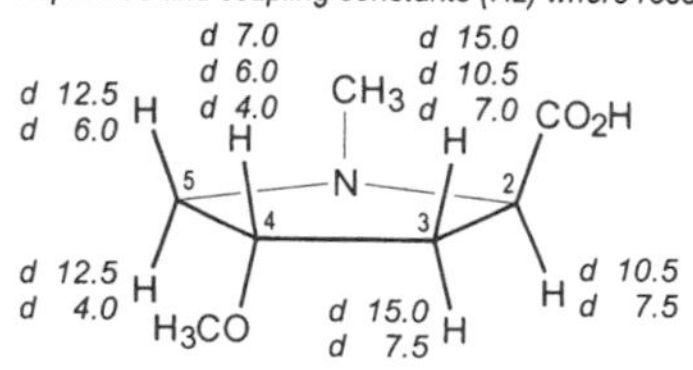

53 Cocaine hydrochloride

First the five protons (integral) of the ^1H NMR spectrum (δ_H = 7.50 - 7.94) in the chemical shift range appropriate for aromatics indicate a monosubstituted benzene ring with typical coupling constants (8.0 Hz for *ortho* protons, 1.5 Hz for *meta* protons.). The chemical shift values especially for the protons which are positioned *ortho* to the substituent (δ_H = 7.94) reflect a –M effect. Using the CH COLOC plot it can be established from the correlation signal δ_C/δ_H = 166.4/7.94 that it is a benzoyl group **A**.

In the HH COSY plot it is possible to take as starting point the peripheral H signal at δ_H = 5.59 in order to trace out the connectivities **B** of the aliphatic H atoms:

B δ_H 4.07 → 2.44 → 5.59 → 3.56 → 4.27 → 2.51

It is then possible to read off from the CH COSY plot those CH links **C** which belong to **B** and to see that between δ_H = 2.22 and 2.51 the protons of approximately three methylene groups overlap (integral). Two of these form AB systems in the ^1H NMR spectrum (δ_H = 2.24 AB 2.44 at δ_C 23.7; δ_H = 2.22 AB 2.51 at δ_C = 24.9); one pair of the methylene protons approximates the A_2 system (δ_H = 2.44 at δ_C = 33.9) even at 400 MHz.

The ^1H and ^{13}C NMR spectra also indicate an NCH_3 group (δ_C/δ_H = 39.6/2.92) and an OCH_3 group (δ_C/δ_H = 53.4/3.66). The CH connectivities **D** of the NCH_3 protons (δ_H = 2.92) across three bonds to the C atoms at δ_C = 65.3 and 64.5, derived from the CH COLOC plot, are especially informative, because the combination of **C** and **D** gives the N-methylpiperidine residue **E** with four spare bonds:

The CH COLOC diagram also shows

■   the linkage **F** of the OCH_3 protons (δ_H = 3.66) with the carboxy C atom at δ_C = 174.1,

■   the connection **G** of the proton at δ_H = 5.59 with the same carboxy C atom,

■   and the CH fragments **H** and **I** involving the protons at δ_H = 4.07 and 4.27;

if **B** and **C** are taken into account then the coupling partners (δ_C/δ_H = 24.9 and 4.07 and δ_C/δ_H = 23.7 and 4.27) must be separated by three bonds.

(δ_C : upright , δ_H : italics)

Thus the ecgonin methyl ester fragment **J** can clearly be recognised; only the link to the *O* atom still remains to be established. The attachment of the *O* atom is identified by the large chemical shift value ($\delta_H = 5.59$) of the proton on the same carbon atom. The parts **A** and **J** then give the skeleton **K** of cocaine.

The fine structure of the ^1H signal at $\delta_H = 5.59$ (3-*H*) reveals the relative configuration of C-2 and C-3. A doublet (*11.5 Hz*) of pseudotriplets (*7.0 Hz*) is observed for an *antiperiplanar* proton (*11.5 Hz*) and two *synclinal* coupling partners (*7.0 Hz*). From that the *cis* configuration of benzoyloxy- and methoxycarbonyl groups is deduced (structure **L**).

The orientation of the NC*H₃* group, whether *syn* or *anti* to the methoxycarbonyl function, is shown by the NOE difference spectrum in tetradeuteriomethanol. If the *N*-methyl proton ($\delta_H = 2.92$) is decoupled an NOE effect is observed for the protons at $\delta_H = 4.27$, *4.07* and between $\delta_H = 2.44$ and *2.51* but not merely at *2.44*. Thus, in tetradeuteriomethanol the *N*-methyl group is positioned *anti* to the methoxy carbonyl group. Hence the assignment of the *endo* and *exo* protons on C-6 and C-7 in the structure **M** of cocaine hydrochloride can also be established.

(δ_C : upright , δ_H : italics)

CH multiplicities , CH coupling constants (Hz) , coupling protons

C-1	D	155.5				
C-2	D	141.1				
C-3	D	153.6	d* 7.9 (1-H)	d* 7.9 (5-H)	d* 7.9 (4-H)	* "q"
C-4	T	133.9				
C-5	D	155.5				
C-6	T	135.9				
C-7	T	135.9				
NCH₃	Q	143.7				
2-COO	S		b			
OCH₃	Q	147.7				
C-1′	S		t 7.9 (3′,5′-H₂)			
C-2′,6′	D	163.4	d* 5.9 (6′-/2′-H)	d* 5.9 (4′-H)		* "t"
C-3′,5′	D	163.4	d 7.9 (5′-/3′-H)			
C-4′	D	161.4	d 7.9 (2′,6′-H₂)			
1′-COO	S					

M
Cocaine hydrochloride

54   Viridifloric acid 7-retronecine ester (heliospathuline)

From the *HH* COSY plot the *HH* relationships **A - D** are read off:

A δ_H 4.43 → 5.64 → $\begin{matrix}1.98^A\\2.14^B\end{matrix}$ → $\begin{matrix}2.60^A\\3.49^B\end{matrix}$ **B** δ_H $\begin{matrix}3.37^A\\4.01^B\end{matrix}$ → 5.60 → $\begin{matrix}4.01^A\\4.22^B\end{matrix}$

C δ_H 3.85 → 1.25   **D** δ_H 0.73 → 1.93 → 0.85

These can be completed following interpretation of the *CH* COSY diagram (Table 54.1) to give the structural fragments **A - D**.

A δ_H 4.43 → 5.64 → $\begin{matrix}1.98^A\\2.14^B\end{matrix}$ → $\begin{matrix}2.60^A\\3.49^B\end{matrix}$

δ_C — 76.3 —— 76.6 —— 34.8 —— 53.9 —
         —CH——— CH——— CH₂——— CH₂—

B δ_H $\begin{matrix}3.37^A\\4.01^B\end{matrix}$ → 5.60 → $\begin{matrix}4.01^A\\4.22^B\end{matrix}$

δ_C — 62.5 —— 124.3 = 139.1 — 59.4 —
         —CH——— CH=——— C ——— CH₂—

C δ_H 3.85 → 1.25

δ_C — 72.5 — 16.6
         —CH——— CH₃

D δ_H 0.73   1.93   0.85

δ_C 17.2 — 31.9 — 15.7
         H₃C——— CH——— CH₃

The molecular formula contains four double-bond equivalents, of which the ^{13}C NMR spectrum identifies a carboxy group ($\delta_C = 174.4$) and a CC double bond ($\delta_C = 139.1$: C, and $\delta_C = 124.3$: *CH* with *H* at $\delta_H = 5.60$ from *CH* COSY) on the basis of the three signals in the sp^2 chemical shift range. The two additional double-bond equivalents must therefore belong to two separate or fused rings. Since fragments **A** and **B** terminate in electronegative heteroatoms judging from their ^{13}C ($\delta_C = 62.5, 53.9$ and 76.3) and ^1H chemical shift values (Table 54.1), a pyrrolizidine bicyclic system **E** is suggested as the alkaloid skeleton, in line with the chemotaxonomy of *Heliotropium* species, in which fragments **A** and **B** are emphasised with bold lines for clarity.

Table 54.1. Intepretation of the *CH* COSY and the *CH* COLOC plots and the DEPT subspectra

Partial structure	Protons δ_H	δ_C	C atoms separated by one bond $CH_n{}^a$	two or three bonds δ_C	δ_C
A	5.64	76.6	CH	174.4	76.3
B	5.60	124.3	CH		
A	4.43	76.3	CH		
B	4.01 AB 4.22	59.4	CH_2	139.1	
C	3.85	72.5	CH		
B	3.37 AB 4.01	62.5	CH_2	139.1	124.3
A	2.60 AB 3.49	53.9	CH_2		
A	1.98 AB 2.14	34.8	CH_2		
D	1.93	31.9	CH		
C	1.25	16.6	CH_3		
D	0.85	15.7	CH_3		
D	0.73	17.2	CH_3	83.9	31.9
	H atoms bonded to C : H_{22}			(therefore 3 O*H*)	

^{a} *CH* multiplicities from the DEPT ^{13}C NMR subspectra.

Correlation signals of the *AB* systems $\delta_H = 4.01^A 4.22^B$ and $3.37^A 4.01^B$ with the C atoms of the double bond (at $\delta_C = 124.3$ and 139.1, Table 54.1) confirm the structural fragment **B**. A signal relating the proton at $\delta_H = 5.64$ (Table 54.1) to the carboxy C atom (at $\delta_C = 174.4$) shows that the O*H* group at C-7 has esterified (partial formula **F**) in accord with the higher ^1H shift (at $\delta_H = 5.64$) of proton 7-*H* caused by the carboxylate. When the O*H* group at C-7 is unsubstituted as in heliotrin then 7-*H* appears at $\delta_H = 4.06$.[31] On the other hand, the chemical shift values of the *AB* protons at C-9, which are considerably lower than those of heliotrin, indicate that the 9-O*H* group is not esterified.

The relative configuration at C-7 and C-8 cannot be established from the *HH* coupling constants; for five-membered rings the relationships between dihedral angles and coupling constants for *cis* and *trans* configurations are not as clear as for six-membered rings. However, NOE difference spectra shed light on the situation: decoupling at $\delta_H = 5.64$ (7-*H*) leads to a very distinct NOE at $\delta_H = 4.43$ (8-*H*) and vice versa. The protons 7-*H* and 8-*H* must therefore be positioned *cis* (**G**). Decoupling of 7-*H* also leads to an NOE on the protons 6-H^A and 6-H^B, which indicates the spatial proximity of these three protons. A Dreiding model shows that the envelope conformation of the pyrrolidine ring (C-7 lies out of the plane of C-8, *N*, C-5 and C-6), in fact places 7-*H* between 6-H^A and 6-H^B so that the distances to these protons do not differ substantially. The 7-*H* signal splits accordingly into a pseudotriplet with *3.5 Hz*; 8-*H* and 6-H^A are coupling partners of 7-*H* (dihedral angle ca 60°), whilst 6-H^B and 7-*H* have a dihedral angle of 90° so no more couplings are detected.

Finally, fragments **C** and **D** belong to the acidic residues in the alkaloid ester. Taking into account the two *OH* groups (cf. Table 54.1), the *CH* correlation signal of the methyl protons at $\delta_H = 0.73$ with the quaternary C atoms at $\delta_C = 83.9$ links **C** and **D** to the diastereomers viridifloric or trachelanthic acid; distinction between the two is discussed in more detail in the literature [31]. The diastereotopism of the isopropyl methyl C atoms is a good criterion for making the distinction. Their chemical shift difference is found to be $\Delta\delta_C = 17.2 - 15.7 = 1.5$, much closer to the values reported for viridifloric acid ester ($\Delta\delta_C \approx 1.85$; for trachelanthic acid ester a value of $\Delta\delta_C \leq 0.35$ would be expected). Thus structure **H** of the pyrrolizidine alkaloid is established. It can be described as viridifloric acid-7-retronecine ester or, because of its plant origin, as heliospathuline [45].

H

(δ_C : upright , δ_H : *italics*)

55 Actinomycin-D

2-Amino-4,6-dimethyl-3-oxo-3*H*-phenoxazine-1,9-dicarboxylic acid also named actinocin is the chromophor of the red antineoplastic chromopeptide actinomycin D (formula **A**). Two cyclopentapeptide lactone rings (amino acids: L-threonine, D-valine, L-proline, sarcosine, and *N*-methyl-L-valine) are attached to the carboxy carbons of actinocin by two amide bonds involving the amino groups of threonine.

A
Actinomycin D
Sar = Sarkosine (*N*-Methylglycine)
MeVal = *N*-Methylvaline

The attachment of the cyclopentapeptide lactone rings to the carboxy functions at C-1 and C-9 of the actinocin heterocycle **B** can be deduced from the HMBC plot **d**: Protons 7-H ($\delta_H = 7.35$) and 8-H ($\delta_H = 7.62$) of the heterocycle display an *AB* system in the proton domain. The attached carbon nuclei have been assigned by *CH* COSY in the literature ($\delta_H/\delta_C = 7.35/130.33$ and 7.62/125.93, ref. [6], p. 426); the other carbon atoms of the benzenoid ring within the phenoxazone are assigned as reported [6] by correlation via $^2J_{CH}$ and $^3J_{CH}$ coupling detected in **d** which additionally shows a weak $^4J_{CH}$ correlation signal of 8-H with C-5a ($\delta_H/\delta_C = 7.62/140.53$).

Proton 8-H ($\delta_H = 7.62$) gives a correlation signal via $^3J_{CH}$ coupling in the HMBC plot **d** with the carboxy carbon at $\delta_C = 166.18$ which also correlates with the N*H* proton of threonine at $\delta_H = 7.15$. via $^2J_{CH}$. Starting with this proton, all protons in the cyclopentapeptide lactone ring attached to C-9 are assigned by means of the spatial correlation signals between the amino acid protons in the *HH* ROESY experiment (**c**) which occur in addition to the proton connectivities detected for the individual amino acids in the *HH* COSY and *HH* TOCSY (**a** and **b**). The latter permit the assignments of proton chemical shifts of all amino acid residues as outlined in Table 55.1, provided they are sufficiently resolved which is not the case for the α-, β- and γ-protons of *N*-methylvaline (α-, β-: $\delta_H = 2.64 - 2.65$, γ: $\delta_H = 0.82$ and 0.94) and the *N*-methyl protons of sarcosine ($\delta_H = 2.85$).

Table 55.1. Evaluation of *HH* COSY and *HH* TOCSY for proton shift (δ_H) assignments of the amino acids

	Thr	Val	Pro	Sar	MeVal
NH	7.15	8.00			
α-H	4.50	3.57	5.93	3.59 AB 4.78	2.64 - 2.65
β-H	5.20	2.20	1.84 AB 2.93		2.64 - 2.65
γ-H	1.24	0.88 , 1.11	2.10 AB 2.26		0.82 , 0.94
δ-H			3.71 AB 3.82		
NCH₃				2.85	2.89
NH	7.76	8.15			
α-H	4.61	3.53	6.00	3.62 AB 4.71	2.64 - 2.65
β-H	5.15	2.13	1.80 AB 2.64		2.64 - 2.65
γ-H	1.25	0.90 , 1.10	2.04 AB 2.28		0.82 , 0.94
δ-H			3.70 AB 3.96		
NCH₃				2.85	2.92

Evaluation of the *HH* ROESY (**c**) verifies the amino acid sequence of the cyclopentapeptide lactone ring attached to C-9 of actinocin by means of the NOE induced spatial correlation signals labelled [a-d] in the formula **C**. The α- and β-protons of threonine in both cyclopentapeptide lactone rings are sufficiently separated but close to each other. This applies to the other amino acid protons with the exception of *N*-methylvaline, in which proton signals overlap (Table 55.1). Therefore, starting from the α- and β-protons of threonine (δ_H = 4.61 and 5.15), the sequence Thr-Val-Pro-Sar-MeVal attached to C-1 of actinocin is similarly verified in the *HH* ROESY as shown in formula **C**. The connection Sar–MeVal is established by the spatial correlations [d] between the *B*-protons of sarcosine and the *N*-methyl-protons of *N*-methylvaline in both rings (δ_H = 4.71/2.92 and 4.78/2.89).

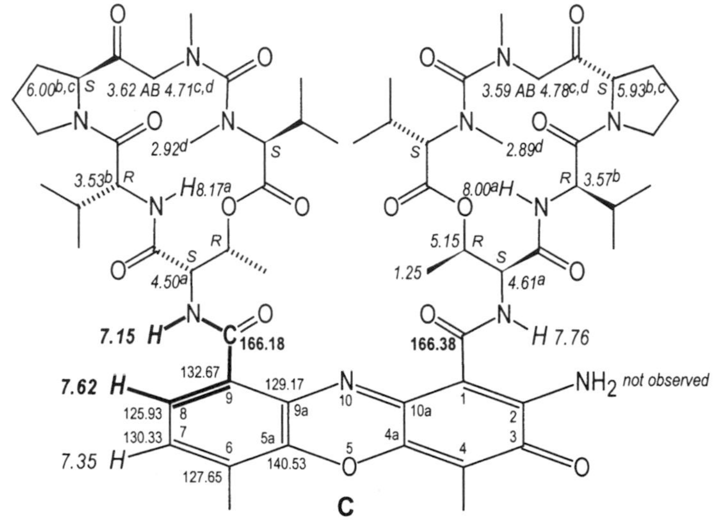

C

The results of Table 55.1 complete the assignment of all proton shifts of both cyclopentapeptide lactones as far as possible as shown in formula **D**.

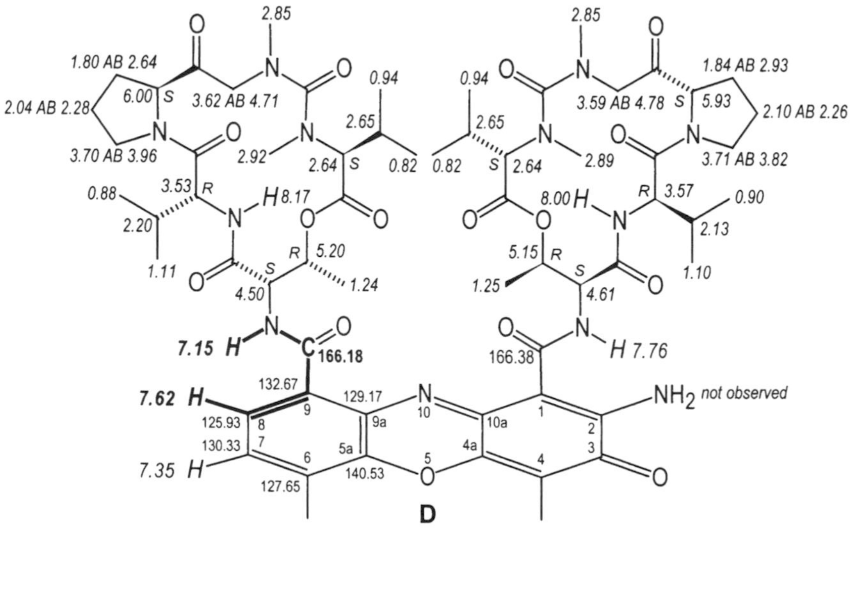

D

Additional significant spatial correlations (NOE) in the *HH* ROESY experiment (**c**) provide information concerning the distances of some protons from one ring (at C-9) to the other (at C-1). Such closely spaced protons are:

Pro-α *(6.00)* – Thr-α *(4.61)* ; Pro-α *(5.93)* – Thr-α *(4.50)*

Thr-α *(4.61)* – Pro-δ^B *(3.96)* ; Thr-α *(4.50)* – Pro-δ^B *(3.82)*

Pro-δ *(3.96)* – Thr-CH_3 *(1.25)* ; Pro-δ *(3.82)* – Thr-CH_3 *(1.24)*

Thr-N*H* *(7.76)* – Thr-NH *(7.15)*

Molecular modelling calculations using values of smaller than 3 Angstrom units for these proton-proton distances can be performed to obtain an optimized picture of the molecule. But this exceeds the scope of this book.

REFERENCES

1. M. Hesse, H. Meier, B. Zeeh, *Spectroscopic Methods in Organic Chemistry*, Georg Thieme, Stuttgart, New York, 1997.

2. (a) H. Günther, *NMR Spectroscopy*, 3rd edn, Georg Thieme, Stuttgart, New York, 1992.
    (b) M. H. Levitt, *Spin Dynamics – Basic Principles of NMR Spectroscopy*, J. Wiley & Sons, Chichester, 2001.

3. H. Friebolin, *One- and Two-dimensional NMR Spectroscopy. An Introduction*, 3rd edn VCH, Weinheim, 1997.

4. G. C. Levy, R. L. Lichter, G. L. Nelson, *Carbon-13 Nuclear Magnetic Resonance Spectroscopy*, 2nd edn, Wiley-Interscience, New York, 1980.

5. H. O. Kalinowski, S. Berger, S. Braun, *^{13}C NMR Spectroscopy*, J. Wiley & Sons, Chichester, 1988.

6. E. Breitmaier, W. Voelter: *Carbon-13 NMR Spectroscopy – High Resolution Methods and Applications in Organic Chemistry and Biochemistry*, 3rd edn, VCH Weinheim, 1990.

7. G. C. Levy, R. L. Lichter, *Nitrogen-15 Nuclear Magnetic Resonance Spectroscopy*, Wiley-Interscience, New York, 1979.

8. E. Breitmaier, Die Stickstoff-15-Kernresonanz – Grenzen und Möglichkeiten, *Pharm. Unserer Zeit 12* (1983) 161.

9. W. von Philipsborn, R. Müller, ^{15}N NMR Spectroscopy – New Methods and their Applications, *Angew. Chem. Int. Ed. Engl. 25* (1986) 381.

10. C. LeCocq, J. Y. Lallemand, *J. Chem. Soc., Chem. Commun. 1981* 150.

11. D. W. Brown, T. T. Nakashima, D. I. Rabenstein, *J. Magn. Reson. 45* (1981) 302 .

12. A. Bax, *Two-Dimensional Nuclear Magnetic Resonance in Liquids*, Reidel, Dordrecht, 1984.

13. R. R. Ernst, G. Bodenhausen, A. Wokaun, *Principles of Nuclear Magnetic Resonance in One and Two Dimensions*, Oxford University Press, Oxford, 1990.

14. D. M. Dodrell, D. T. Pegg, M. R. Bendall, *J. Magn. Reson. 48* (1982) 323; *J. Chem. Phys. 77* (1982) 2745.

15. M. R. Bendall, D. M. Dodrell, D. T. Pegg, W. E. Hull, *DEPT*; Information brochure with experimental details, Bruker Analytische Messtechnik, Karlsruhe, 1982.

16. J. L. Marshall, *Carbon-Carbon and Carbon-Proton NMR Couplings: Applications to Organic Stereochemistry and Conformational Analysis*, Verlag Chemie International, Deerfield Beach, FL, 1983.

17. (a) J. K. M. Sanders, B. K. Hunter, *Modern NMR Spectroscopy. A Guide for Chemists*, 2nd edn, Oxford University Press, Oxford, 1993. (Introduction to basic one- and two-dimensional NMR experiments and some chemical applications)
    (b) S. Braun, H. O. Kalinowski, S. Berger, *150 and More Basic NMR Experiments, A Practical Course*, 2nd edn, Wiley-VCH, Weinheim, 1998. (Description and setup of many basic and more advanced one- and two-dimensional NMR experiments)

(c) P. Bigler, *NMR Spectroscopy: Processing Strategies*, 2nd edn, Wiley-VCH, Weinheim, 2000. (Introduction to NMR processing also for newcomers)

18. W. Aue, E. Bartholdi, R. R. Ernst, *J. Chem. Phys. 64* (1976) 2229.

19. A. Bax and R. Freeman, *J. Magn. Reson. 42* (1981) 164; *44* (1981) 542.

20. A. Bax, R. Freeman, S. P. Kempsell, *J. Am. Chem. Soc. 102* (1980) 4581.

21. T. H. Mareci, R. Freeman, *J. Magn. Reson. 48* (1982) 158.

22. D. L. Turner, *J. Magn. Reson. 49* (1982) 175.

23. A. Bax, G. Morris, *J. Magn. Reson. 42* (1981) 501.

24. H. Kessler, C. Griesinger, J. Zarbock, H. Loosli, *J. Magn. Reson. 57* (1984) 331 .

25. D. Leibfritz, *Chem. Ber. 108* (1975) 3014.

26. D. Neuhaus, M. P. Williamson, *The Nuclear Overhauser Effect in Structural and Conformational Analysis*, 2nd edn, Wiley-VCH, Weinheim, 2000.

27. M. Kinns, J. K. M. Sanders, *J. Magn. Reson. 56* (1984) 518.

28. G. Bodenhausen, R. R. Ernst, *J. Am. Chem. Soc. 104* (1982) 1304.

29. A. Ejchardt, *Org. Magn. Reson. 9* (1977) 351.

30. J. L. C. Wright, A. G. McInnes, S. Shimizu, D. G. Smith, J. A. Walter, D. Idler, W. Khalil, *Can. J. Chem. 56* (1978) 1898.

31. S. Mohanray, W. Herz, *J. Nat. Prod. 45* (1982) 328.

32. W. H. Pirkle, D. J. Hoover, *Top. Stereochem. 13* (1983) 263.

33. G. R. Sullivan, *Top. Stereochem. 9* (1976) 111.

34. M. Gosmann, B. Franck, *Angew. Chem. Int. Ed. Engl. 25* (1986) 1107; G. Kübel, B. Franck, *Angew. Chem. Int. Ed. Engl. 27* (1988) 1203.

35. H. Kessler, *Angew. Chem. Int. Ed. Engl. 9* (1970) 219.

36. J. Sandström, *Dynamic NMR Spectroscopy*, Academic Press, New York, 1982.

37. M. Oki (Ed.), *Applications of Dynamic NMR Spectroscopy to Organic Chemistry*, VCH, Deerfield Beach, FL, 1985.

38. F. McCapra, A. I. Scott, *Tet. Lett. 1964*, 869 [monordene].

39. H. Kneifel, C. Poszich-Buscher, S. Rittich, E. Breitmaier, *Angew. Chem. Int. Ed. Engl. 30* (1991) 202 [metabolite].

40. S. Damtoft, S. R. Jensen, B. J. Nielsen, *Phytochem. 20* (1981) 2717 [asperuloside].

41. F. Bohlmann, J. Jakupovic, A. Schuster, R. King, H. Robinson, *Phytochem. 23* (1984) 1445; S. Sepúlveda-Boza, E. Breitmaier, *Chem. Ztg. 111* (1987) 187 [hydrocostus acid].

42. E. Graf, M. Alexa, *Planta Med. 1985*, 428 [14-(umbelliferon-7-*O*-yl)driman-3α,8α-diol].

43. M. Garrido, S. Sepúlveda-Boza, R. Hartmann, E. Breitmaier, *Chem. Ztg. 113* (1989) 201 [sendarwine].

44. S. Shibata, O. Tanaka, K. Soma, Y. Iida, T. Ando, H. Nakamura, *Tet. Lett. 1965*, 207; O. Tanaka, S. Yahara, *Phytochem. 17* (1978) 1353 [panaxatriol].

45. E. Röder, E. Breitmaier, H. Birecka, M.W. Fröhlich, A. Badzies-Crombach, *Phytochem. 30* (1991) 1703 [heliospathuline].

FORMULA INDEX OF SOLUTIONS TO PROBLEMS

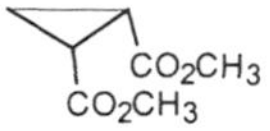

1

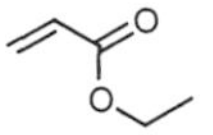

2

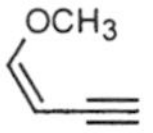

3

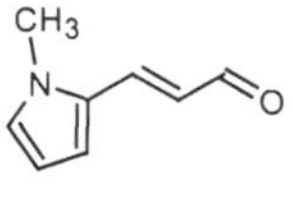

4

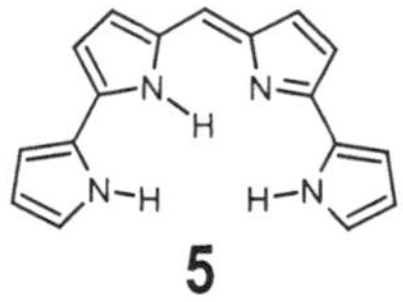

5

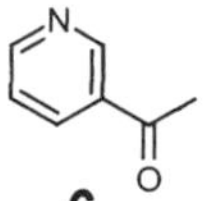

6

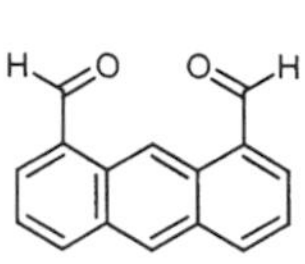

7

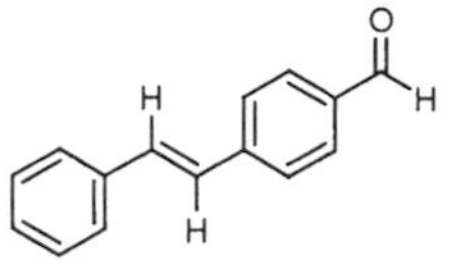

8

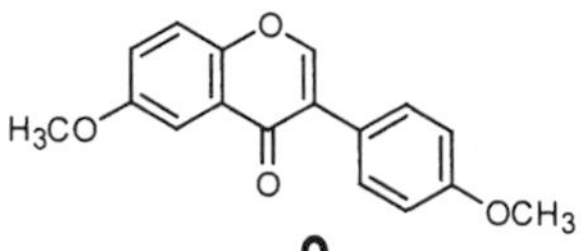

9

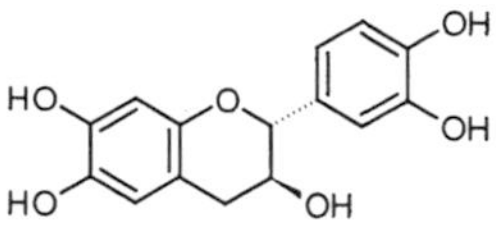

10

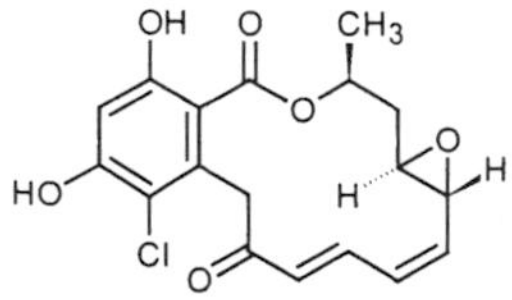

11

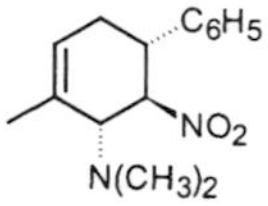

12

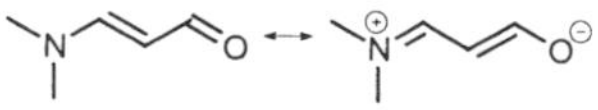

13

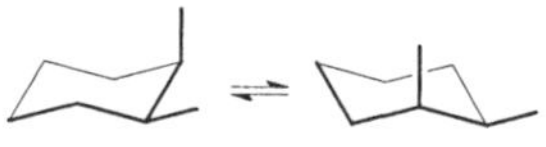

14

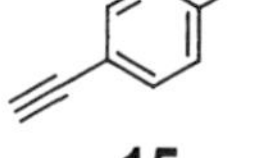

15

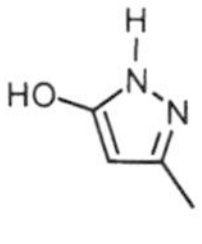

16

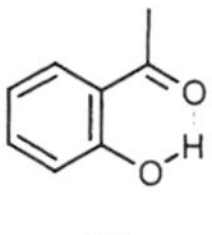

17

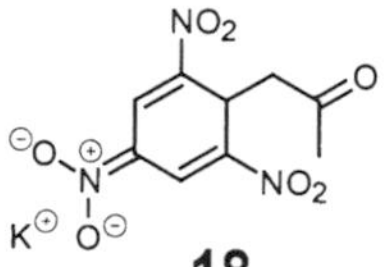

18

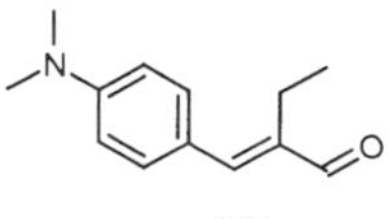

19

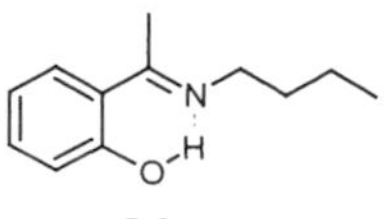

20

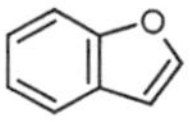

21

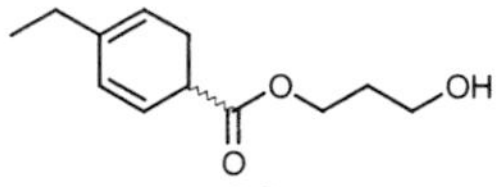

22

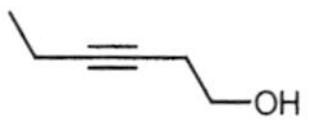

23

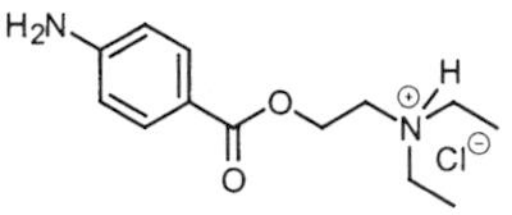

24

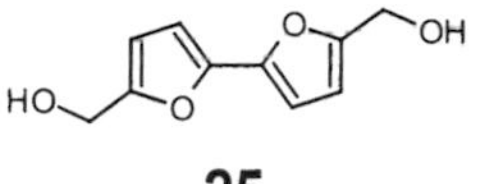

25

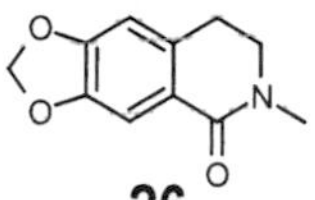

26

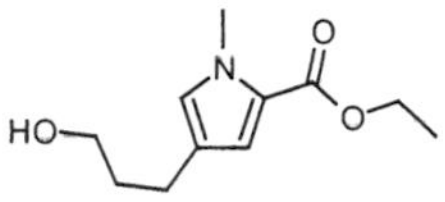

27

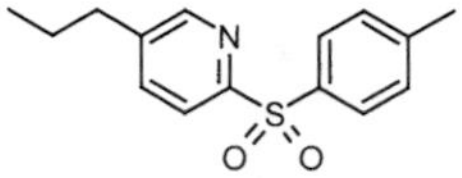

28

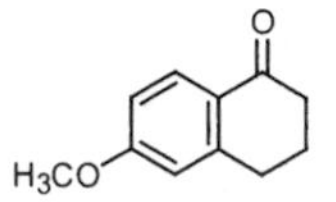

29

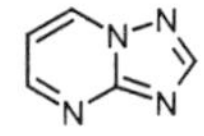

30

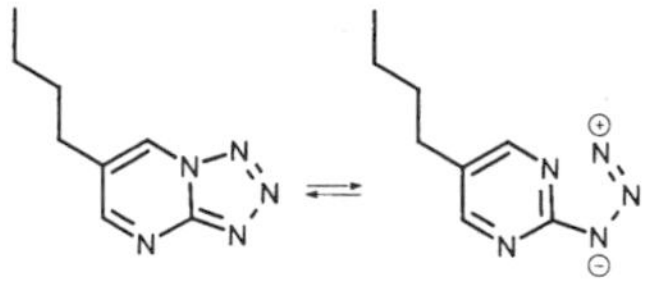

31

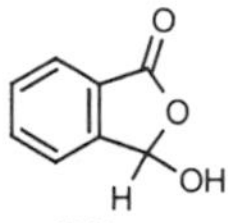

32

33

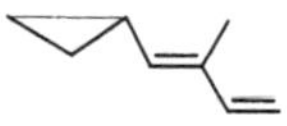

34

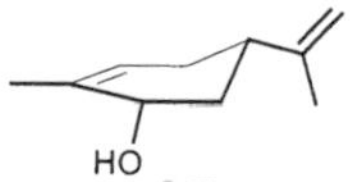

35

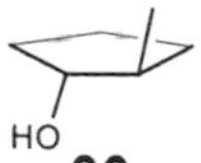

36

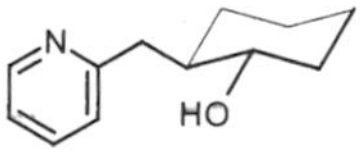

37

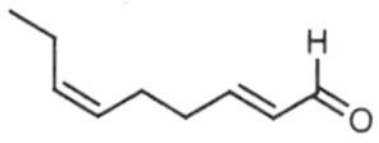

38

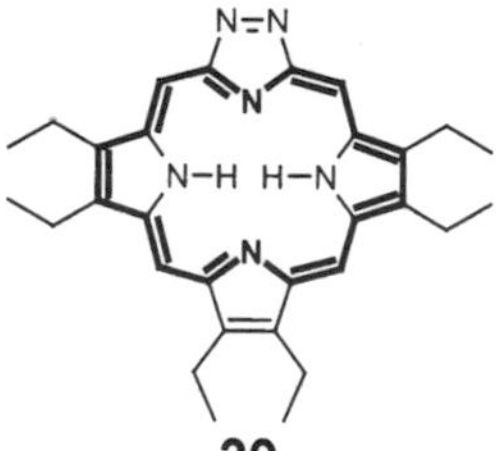

39

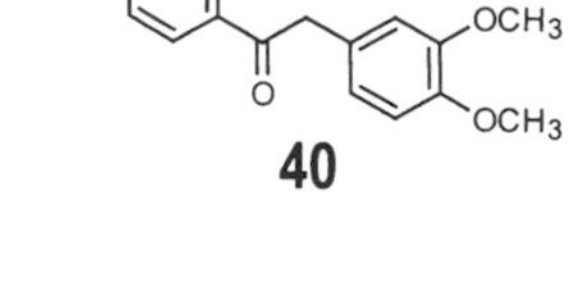

40

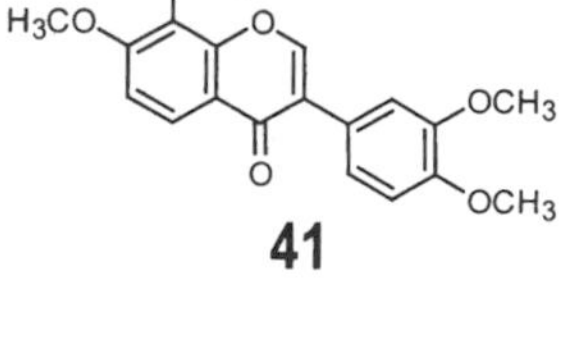

41

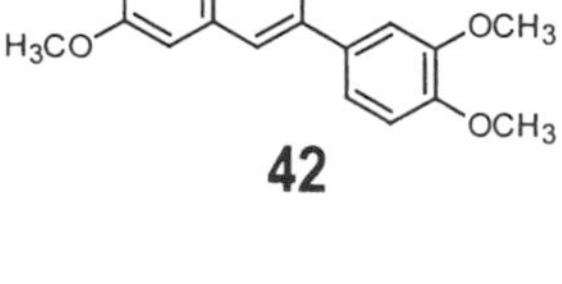

42

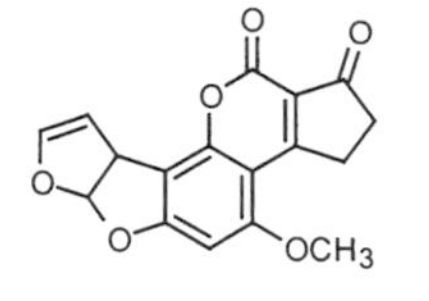

43

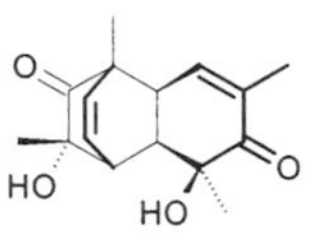

44

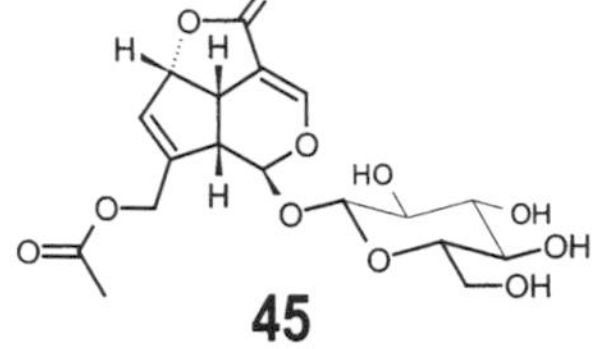

45

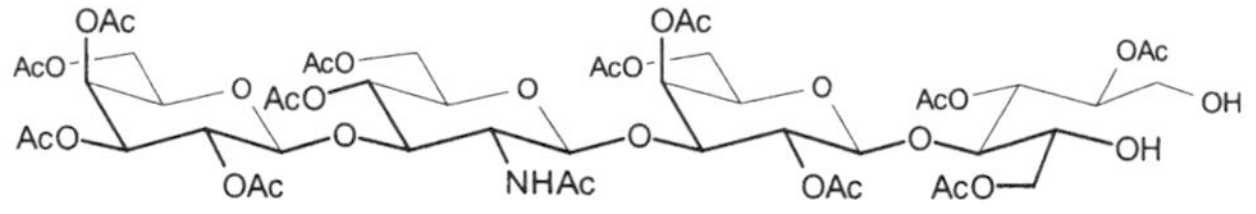

46

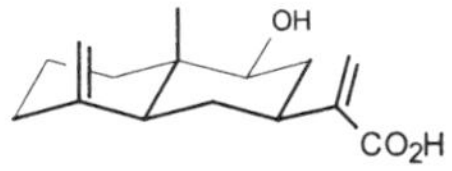

47

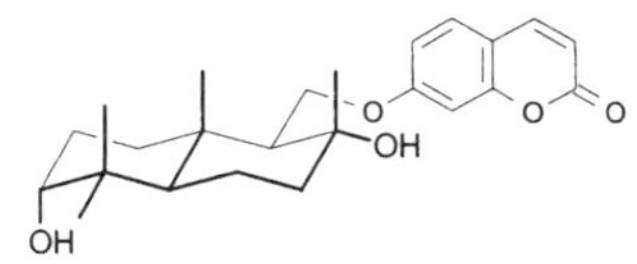

48

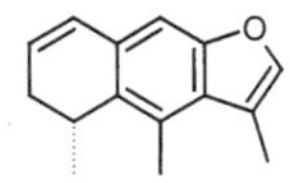

49

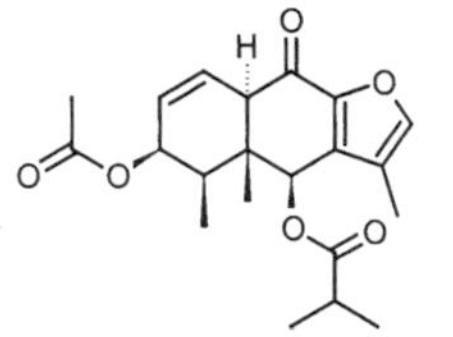

50

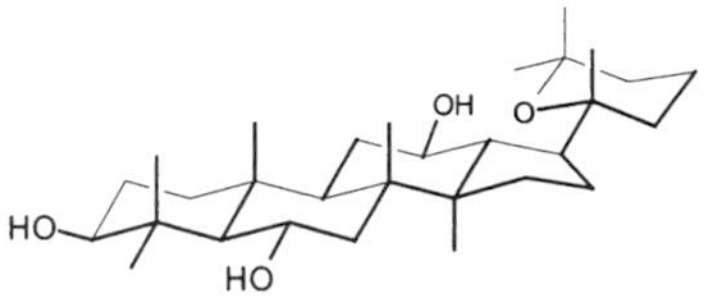

51

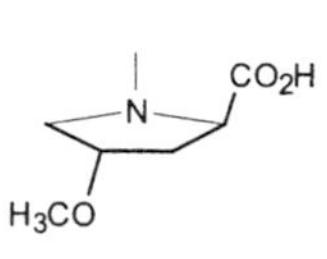

52

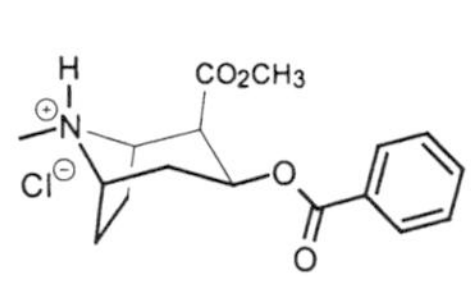

53

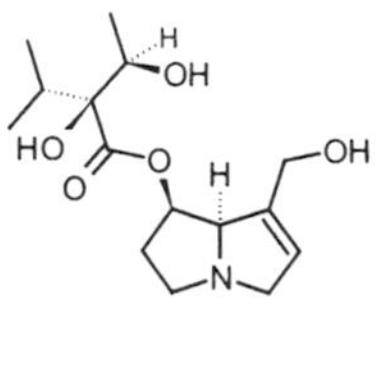

54

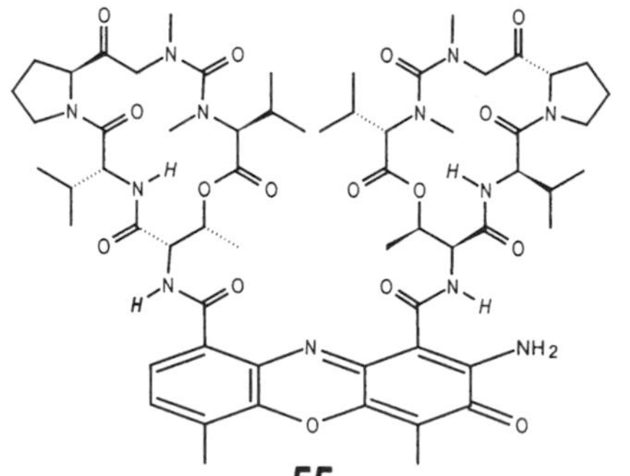

55

SUBJECT INDEX

The abbreviations (**F**, **P**, **T**) are used to imply that the item referred to appears in **F**igures (spectra), in **T**ables or in solutions to **P**roblems.